W9-AOO-561

Small Electrical Appliances and How to Repair Them

New Illustrated Library of Home Improvement Volume 4

Small Electrical Appliances and How to Repair Them

Prentice-Hall/Reston Editorial Staff

Prentice-Hall of Canada, Ltd./Reston Publishing Company
Scarborough, Ontario

Series contributors/ H. Fred Dale, Richard Demske, George R. Drake, Byron W. Maguire, L. Donald Meyers, Gershon Wheeler

*Design/*Peter Maher & Associates
*Color photographs/*Peter Paterson/Photo Design

Printed and bound in Canada

The publishers wish to thank the following companies for
providing photographs for this volume:

John Inglis Co., Ltd.
Orli Metal Products, Ltd.
Proctor Lewyt Division of SCM (Canada), Ltd.
Ronson Products of Canada, Ltd.
Van Wyck Industries (Canada), Ltd.
Westinghouse Canada

Contents

Chapter 4 Small Appliances with Motors 41

Chapter 5 Small Heating Appliances 71

Chapter 6 Preventive Maintenance 113

Small Electrical Appliances and How to Repair Them

1

Fundamentals

1-1 Electricity

Most electrical appliances used in the home are very simple devices and should give many years of trouble-free service. Basically, there are two types: appliances that provide heat, such as toasters, electric irons and percolators; and those that have a motor to provide motion, such as vacuum cleaners, fans and washing machines. Some appliances combine both functions, such as electric clothes dryers, which have a motor to tumble the clothes and a heater to dry them. When an appliance fails to operate correctly, the fault may be mechanical or electrical. Determining the cause of failure is the biggest part of servicing. It is not necessary to be an electrical engineer to do this, but some knowledge of electrical fundamentals is necessary.

To get a better understanding of electricity, it is useful to compare it to the water system in a household. Water reaches the home through large pipes or mains and is under pressure so that it will be forced out of an open faucet which may be many feet higher than the water main. When a tap is opened, a current of water flows in the pipe from the main to the tap. The velocity of the current depends on the diameter of the pipe and also on its smoothness. In general, the larger the diameter and the smoother the pipe, the faster will be the flow of current. However, there is a limit determined by the pressure in the main. Electricity is quite analogous. The water main is replaced by the electric power lines which lead to the house. The electricity is under "pressure". This pressure is called electromotive force (EMF) and is measured in *volts*. When a suitable connection is made to an electrical appliance, current flows, and the speed of this current depends on factors akin to the smoothness and diameter of the water pipe. In general, the larger the diameter of the wire, the faster will be the electric current. The conductivity of the metal used to make the wire is similar to the smoothness of the water pipe. Some metals, such as copper and silver, are excellent conductors. These are said to have low *resistance*, and current flows through copper, for example, much faster than through iron. Nickel and tungsten are poor conductors which tend to slow down the current. The velocity of electrical current is measured in *amperes.*

Let us now assume that a long garden hose is connected to a water faucet. The pressure may be 80 pounds per square inch. If the hose has an inside diameter of about an inch and is quite smooth on the inside, water will flow through it rapidly when the faucet is opened wide. However, if the hose is only one quarter inch in diameter and not smooth, the flow will be restricted. If a larger hose is connected to the open end of the quarter-inch hose, the water will not move faster through this added

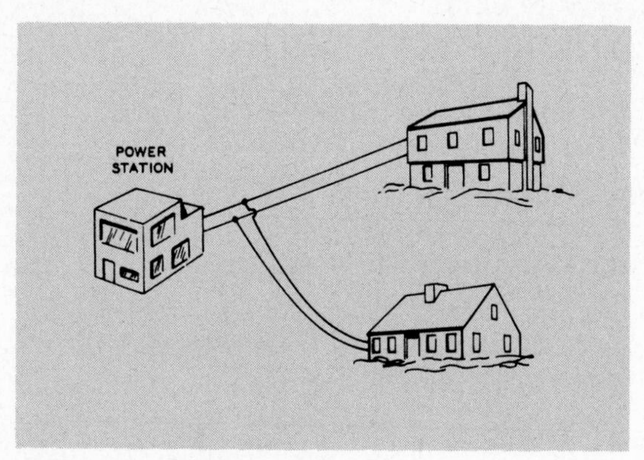

Fig. 1-1. Power distribution.

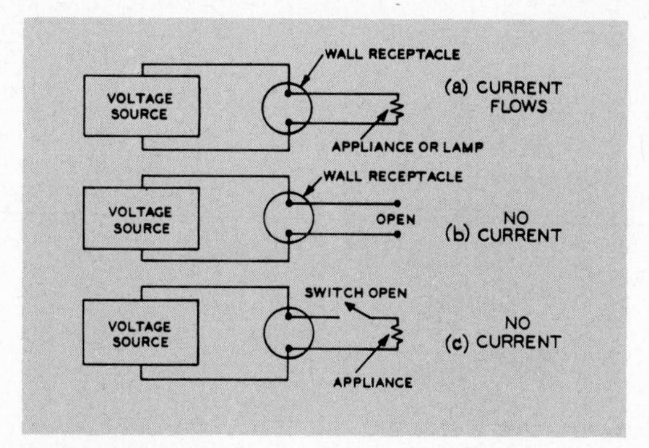

Fig. 1-2. Circuits.

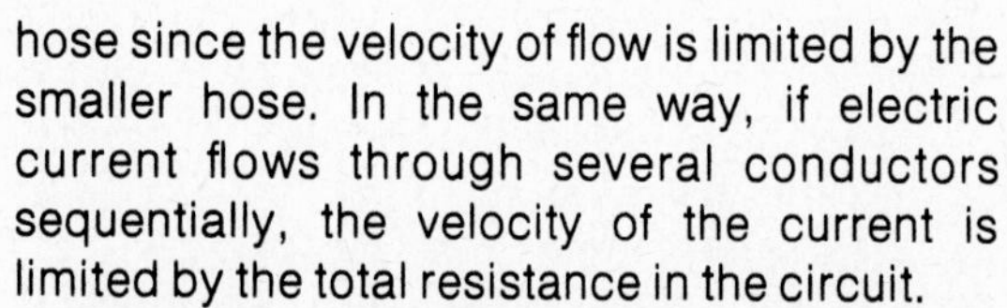

hose since the velocity of flow is limited by the smaller hose. In the same way, if electric current flows through several conductors sequentially, the velocity of the current is limited by the total resistance in the circuit.

In one sense the analogy between electricity and water breaks down completely. If a water faucet is opened, water pours out. But electricity does not pour out of a wall outlet or an empty lamp socket even when the switch is turned on. Thus, electricity or electric current flows only in a *complete circuit.* This is illustrated in Figures 1-1 and 1-2. At a power station a difference in electrical potential (EMF or voltage) is generated. Two wires run from the power station to every house receiving electricity, as shown in Fig. 1-1. These wires go through a distribution center in each home and then to the various electrical outlets in the home such as wall receptacles. In effect, then, each wall receptacle has two terminations which are connected all the way back to the power station or voltage source, as shown in Figure 1-2. When an appliance or lamp is plugged into the receptacle, the circuit is completed and current flows. This is illustrated in (a). The wiggly line labeled "appliance or lamp" is the standard electrical symbol for a resistance and is used here to represent the electrical resistance of the appliance. In (b), the receptacle has nothing plugged into it, and thus no current flows, since the circuit is incomplete. Similarly, in (c), an appliance is plugged into the receptacle, but its switch is in the *off* position. Again, the circuit is incomplete, and no current flows.

The difference in potential, or voltage, between the two terminals in a receptacle is what pushes electricity through a circuit. The two wires or terminals together form a *line*, and each wire by itself is referred to as a *side of the line.* In practice, for safety reasons (which are discussed in the next chapter) and for simplicity and ecomony of installation, one side of the line is always grounded. That is, it is physically attached to the earth. This side is then called the *grounded* side of the line, and the other side is the *hot* side. This means that a voltage exists between the hot side of a line and anything else that is physically attached to the earth. In a home, the cold water pipes are also grounded, that is, attached to earth, and thus a voltage exists between the hot side of the line and the cold water pipes. This is illustrated in Figure 1-3. The plug is removed from the line cord leading to a socket containing a bulb, such as in a table lamp. One wire is attached to the hot side of a receptacle and the other to a faucet or water pipe. The bulb lights. If the wire at the receptacle is moved from the hot side to the ground side, the bulb will not light. Thus, the arrangement shown in Figure 1-3 can also be used to determine which side of the line or receptacle is hot and which is grounded. Similarly, if you touch the hot side of a line and a ground, such as a water pipe, you will get a shock; but you will not get a shock from touching the ground side of the line and another grounded object. You will *not* get a shock even if you touch the hot side, as long as you do not complete the circuit by touching anything else that is

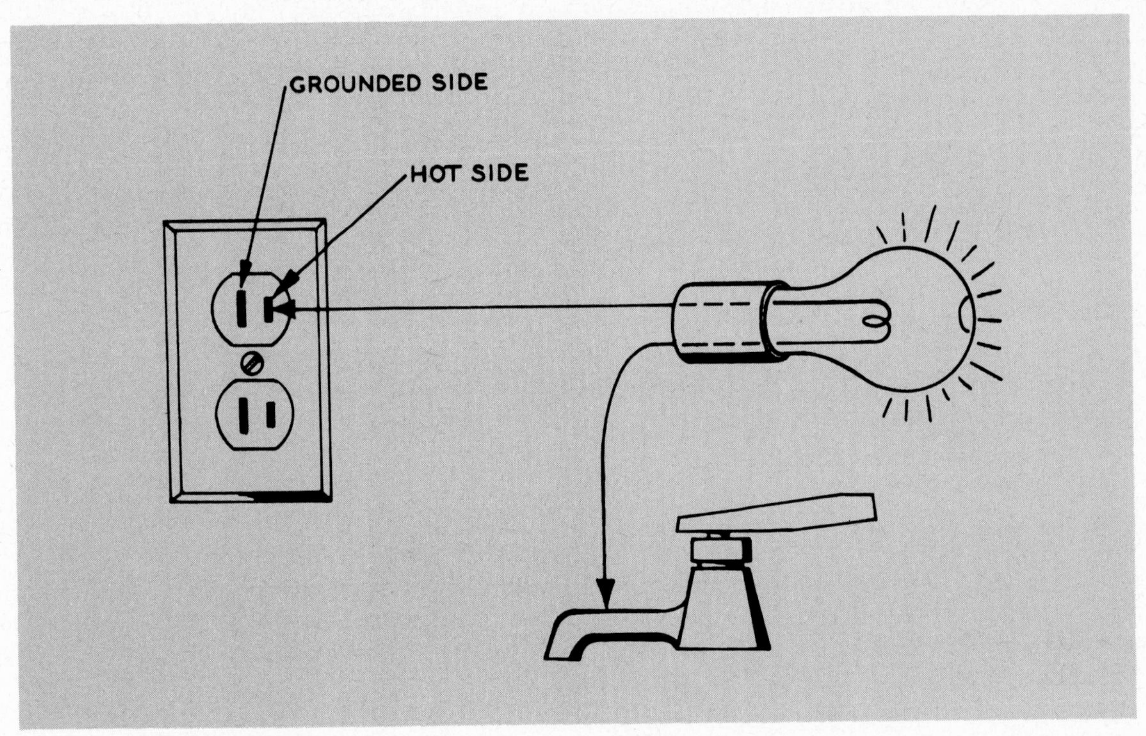

Fig. 1-3. Ground connection.

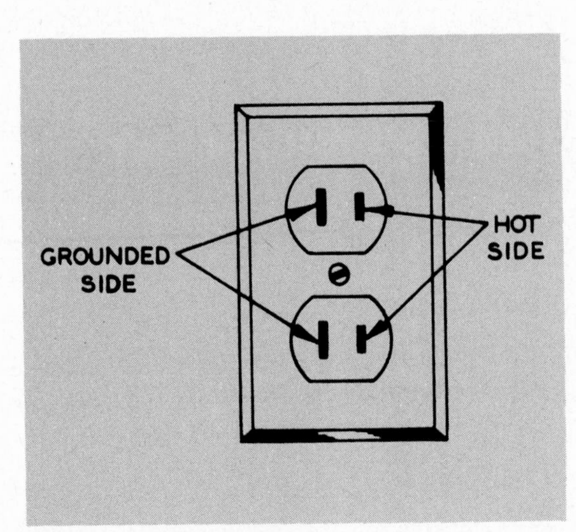

Fig. 1-4. Wall receptacle.

grounded. This means you must not be standing on ground or on a damp floor. This explains why birds are able to light on high voltage wires without electrocuting themselves. They do not complete an electric circuit with any part of their bodies.

If you look carefully at a wall receptacle, you will notice that one of the two slots for a plug is longer than the other, as shown in Figure 1-4. It is common practice to connect the longer slot to the grounded side of the line, as indicated in the figure. The screw in the center, holding the mounting plate to the wall, is also grounded in most installations. Although the slots are of different lengths, the plugs on most lamps and appliances have prongs which are the same width. Thus, a plug can be inserted to connect either side of the appliance line to either side of the house line.

In industrial appliances, for safety reasons, it is desirable to attach the case of the apparatus firmly to ground. This is usually done by using a three-pronged plug and a special receptacle. The receptacle has a third hole which is round and connected directly to ground, as shown in Figure 1-5. A wire from the case of the appliance is connected to the round prong on the plug so that the case is grounded when the appliance is plugged into the receptacle.

Fig. 1-5. Wall receptacle with ground connection.

When electrical current flows through a resistance, it is converted to some other form of energy. This may be heat, light, or mechanical motion, or some combination of these. Thus, when current flows through the filament in an electric bulb, it is converted to light, and when current flows through the heating element in a toaster, it is converted to heat. However, the bulb also gets hot, and the heating element glows. In every electrical apparatus, some of the electrical energy is converted to unwanted heat or light.

The unit of electrical current is the *ampere.* This is a unit of velocity indicating flow of electrical charges per second, just as water current is measured in feet per second. The amount of electrical *power* consumed by an appliance is measured in *watts.* For appliances which produce light or heat, the power consumed in watts is equal to the *product* of the *current* times the *voltage.* For example, one model of electric toaster is rated at 115 volts, 1320 watts. This means that in operation the toaster will draw 1320/115 or 11.5 amperes. Appliances which have motors that produce motion usually use less power than the product of current times voltage. For example, one model of juicer draws 1.75 amperes at 115 volts. The product of 1.75 times 115 is 201.25, but the juicer actually uses only 125 watts.

You pay for electrical energy on the basis of the amount of power used and the time it is used. The unit you pay for is the *kilowatt-hour* (Kwh). A kilowatt is 1000 watts and kilowatt-hours are the product of kilowatts and hours. Thus, if you used the 1320-watt toaster for one quarter hour and the 125-watt juicer for one fifth hour, you would use $1/4 \times 1320$ plus $1/5 \times 125$ or $330+25=355$ watt-hours = 0.355 Kwh. Note that if a 50-watt bulb is left on for ten hours, it uses only half a Kwh.

When electric current flows through copper wire, there is some power dissipated, even though the wire is an excellent conductor with very low resistance. The greater the current for a fixed diameter of wire, the greater will be the power absorbed by the wire. This power loss becomes evident as heat. When too great a current flows through a wire, the heat generated may burn the insulation and cause a fire. The safe current-handling capacity of a copper wire depends on the diameter of the wire. Thus, the choice of wire size, not only for house wiring, but also for line cords for individual appliances, is affected by the expected current.

The electrical system in a home consists of a service line from the main power lines to the house, a distribution box where the service line divides into many branch lines going to separate rooms, and finally the receptacles and sockets to which the electrical appliances are connected. The usual voltage for most electrical appliances is nominally 115 volts, although they will work well at any value between 110 and 120 volts. However, some modern high-power appliances, such as electric ranges and electric clothes dryers require twice this value to furnish enough power to supply sufficient heat. (Recall that the power is the product of volts times amperes.) To supply both 115 volts and 230 volts, a three-wire service line connects the main power line to the house. This line, shown in Figure 1-6, has one wire grounded; each of the other wires is at a potential of 115 volts. Alternating current (AC) is used, and the voltages on the two hot lines are 180° or completely out of phase. This means that when one is 115 volts above ground, the other is 115 volts below, and the difference between them is 230 volts. This last voltage is used to operate a range, clothes dryer and

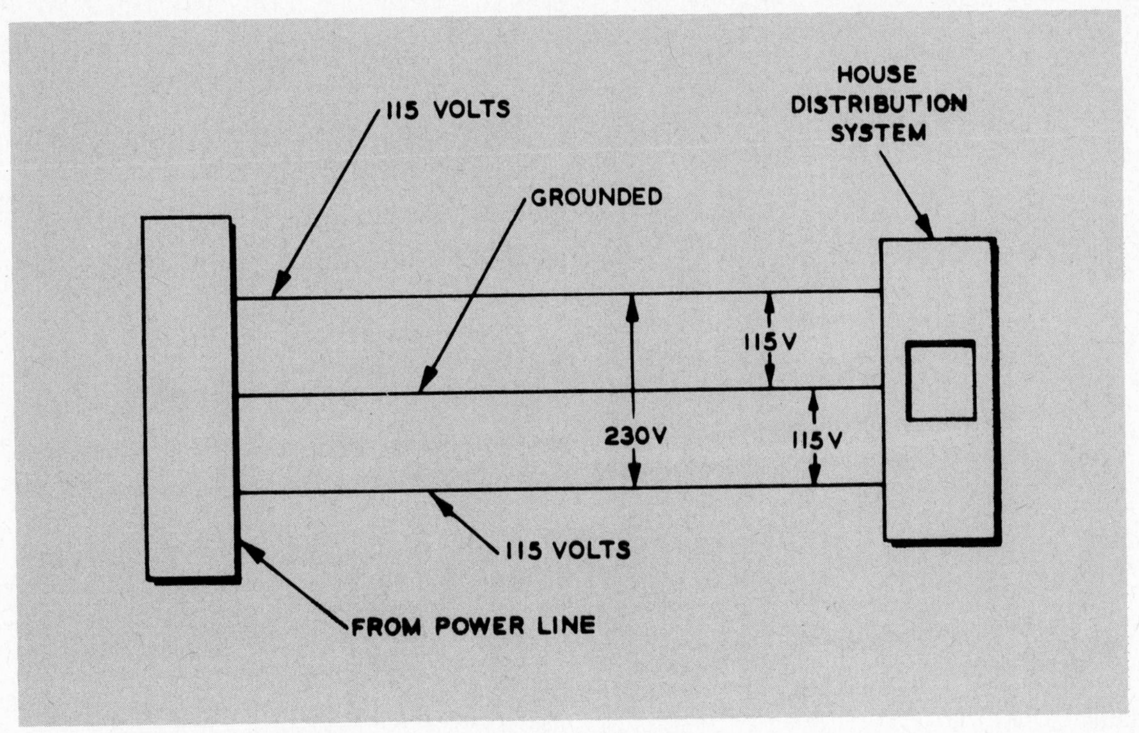

Fig. 1-6. Three-wire service line.

other high-voltage appliances, while small appliances are operated between either hot wire and ground.

Before considering the current requirements of house wiring, it is necessary to know what current is drawn by each appliance. Typical values of current for various appliances are shown in Table 1-1. Individual appliances may have slightly different requirements, but in general those appliances which supply much heat or do heavy work draw greater currents. A typical household may have 20 or 30 appliances, but fortunately only a few are used at any one time. The exact value of current drawn by an appliance or information from which this value can be determined is supplied either on nameplates affixed to the appliance, or from information engraved on the appliance itself. Some typical nameplates are shown in Figure 1-7. Each nameplate carries at least the following information: the name and address of the manufacturer, the model number of the appliance, the voltage required (usually 110 to 120 volts), and either the power consumed or

Table 1-1. Typical Current Requirements.

Lighting	
Ceiling or wall (each bulb)	40-150 w
Floor lamps (each)	150-300 w
Fluorescent lights (each tube)	15-40 w
Pin-to-wall lamps	50-150 w
Table lamps (each)	50-150 w
Ultraviolet lamp	385 w
Lighting and general purpose load for a 1500 sq ft home (1500 sq ft x 4 watts/sq ft)	6000 w
Appliances	
Baker (portable)	800-1000 w
Bed cover	200 w
Bottle warmer	95 w
Broiler-rotisserie	1320-1650 w
Built-in oven	4000 w
Built-in range units	3300 w

Casserole	1350 w
Clock	2 w
Coffeemaker or percolator	440-1000 w
Coffee grinder	150 w
Corn popper	1350 w
Deep fat fryer	1350 w
Egg cooker	500 w
Fan (portable)	100 w
Food blender	230-250 w
Hair dryer	235 w
Hand iron (steam or dry)	1000 w
Heating pad	60 w
Heated tray	500 w
Ice cream freezer	115 w
Ironer	1650 w
Knife sharpener	100 w
Lawn mower	250 w
Mixer	100 w
Portable heater	1000 w
Radio (each)	100 w
Record changer	75 w
Refrigerator*	150-250 w
Roaster	1650 w
Sandwich grill	1200 w
Saucepan	1000 w
Sewing machine	75 w
Shaver	12 w
Skillet	1200 w
Television	300 w
Toaster (modern automatic)	1200 w
Vacuum cleaner	125 w
Ventilating fan (built-in)	140 w
Waffle baker	1200 w
Warmer (rolls, etc)	100 w
Waxer-polisher	350 w

*Each time the refrigerator starts, it momentarily uses several times this wattage.

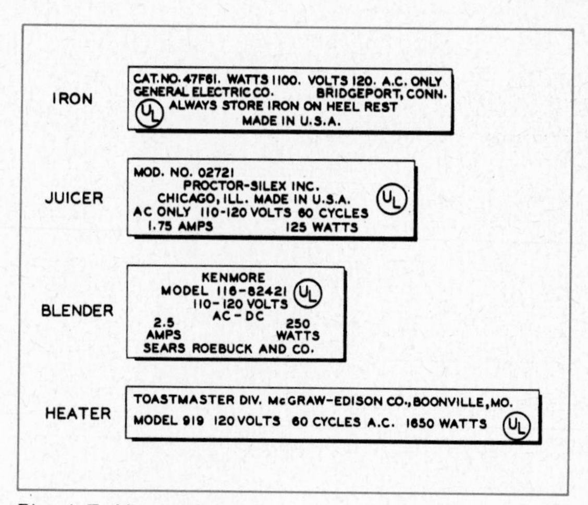

Fig. 1-7. Nameplates.

the current drawn or both. In the United States, the symbol *UL* in a circle shows that the appliance meets requirements of the Underwriter's Laboratory. In Canada, the letters CSA on the nameplate mean Canadian Standards Approved. In appliances that supply only heat or light, the current in amperes is equal to the power in watts divided by the voltage. This is another way of saying that volts times amperes equals watts. Therefore, from the nameplates in Figure 1-7, it is possible to determine that the iron draws 1100/120 or about nine amperes and the heater draws 1650/120 or about 13.8 amperes. The currents for the juicer and blender are specified at 1.75 and 2.5 amperes, respectively, because in appliances with motors the current drawn is usually greater than watts divided by volts.

The nameplate may have additional information as to proper use or care of the appliance. Thus, on the plate for the iron in Figure 1-7, there is a direction for proper storage of the iron. This is to prevent scratching the face of the iron when it is not in use. Toasters usually have plates with additional information for two reasons. First, the toaster is a small appliance which draws more current than many larger ones; and second, the exposed wires in a toaster are a potential shock hazard. Nameplates for two models of toaster are shown in Figure 1-8. Note that

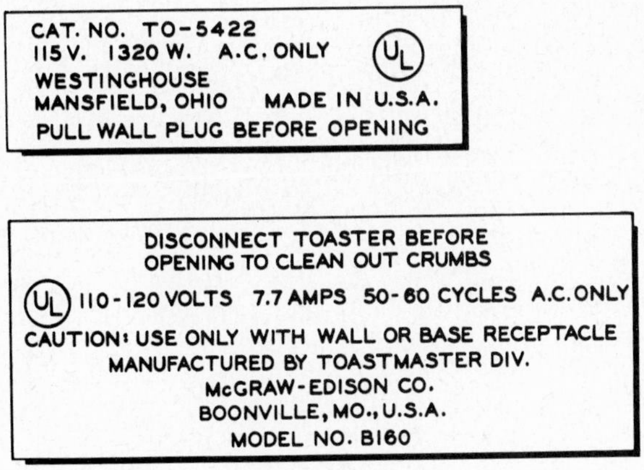

CAT. NO. TO-5422
115 V. 1320 W. A.C. ONLY (UL)
WESTINGHOUSE
MANSFIELD, OHIO MADE IN U.S.A.
PULL WALL PLUG BEFORE OPENING

DISCONNECT TOASTER BEFORE
OPENING TO CLEAN OUT CRUMBS
(UL) 110-120 VOLTS 7.7 AMPS 50-60 CYCLES A.C. ONLY
CAUTION: USE ONLY WITH WALL OR BASE RECEPTACLE
MANUFACTURED BY TOASTMASTER DIV.
McGRAW-EDISON CO.
BOONVILLE, MO., U.S.A.
MODEL NO. B160

Fig. 1-8. Nameplates for toasters.

both warn the user to "pull the plug" before opening the unit to clean out crumbs.

From the information in the Table, it is possible to "design" the house wiring for a typical home. One possible arrangement is shown in Figure 1-9. Each line in this figure represents a two- or three-wire electrical line. A three-wire service line connects the distribution box to the main power line. The service line is rated for 100 amperes and has a 100-ampere fuse or circuit breaker in series with it. The service line also passes through the electric meter before entering the box so that the amount of energy used may be determined. Several branch circuits emerge from the distribution box, and these are rated and fused according to their use. The *lighting* circuits are rated at 15 amperes each. These are wired throughout the house for lamps, small fans, electric razors, toothbrushes, radios and TV sets. One branch line may service two or three rooms. A better arrangement is to have one branch for overhead lights and one or more for wall outlets in living room and bedrooms. Since heavier loads are required in the kitchen, the *appliance* circuits leading there are rated at 20 amperes each. The overhead lights in the kitchen should be on one of the 15-ampere circuits. On the nameplate shown for the Toastmaster toaster in Figure 1-8, the "Caution: Use with wall or base receptacle" is to ensure that the toaster will be plugged into a 20-ampere appliance circuit rather than a 15-ampere lighting circuit. The 230-volt branches are used for high-power equipment, and a separate

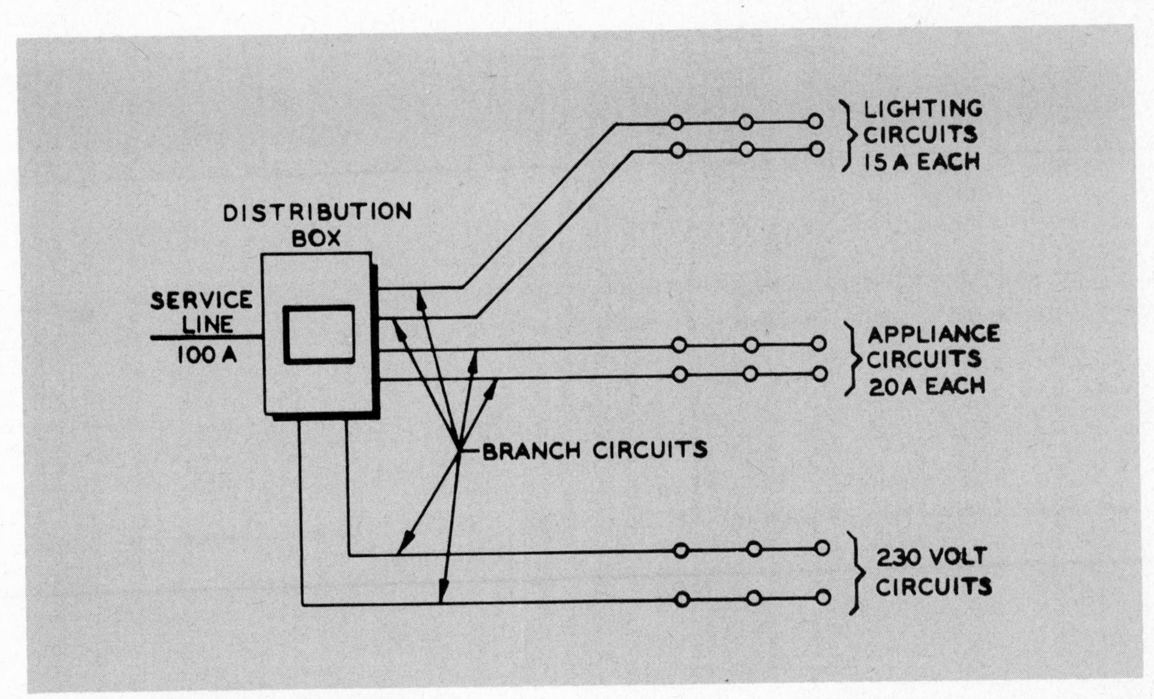

Fig. 1-9. Typical wiring arrangement.

Table 1-2. Some Appliances Requiring Individual or Special Purpose Circuits

Appliance	Wattage
Automatic washer	700 w
Built-in bathroom heater (each)	1000-1500 w
Dishwasher	1200 w
Dishwasher-waste disposer	1500 w
Central air-conditioning system	5000 w
Clothes dryer	5000 w
Clothes dryer (high-speed)*	9000 w
Home freezer	350 w
Mechanism for fuel-fired heating plant	800 w
Range (single oven)	14,000-16,000 w
Room air conditioner (3/4 ton, each)	1200 w
Water heater	2000-4000 w
Water heater (quick-recovery)*	9000 w
Water pump	700 w
Waste disposer alone (without dishwasher)	500 w

* If the high-speed electric clothes dryer or the quick-recovery water heater is to be installed, it is recommended that more than 100-ampere capacity be provided in the electric service entrance, as either of these appliances, in combination with other equipment, would quickly absorb the total available wattage.

branch should be used for each 230-volt appliance. These would include the clothes dryer, air conditioner and range.

If too many appliances are in operation on a single circuit simultaneously, the total current drawn can exceed the safe rating of the wire. To prevent overheating and possible fires from excessive current, each branch has a fuse or circuit breaker which is designed to open the circuit when the rated current is exceeded. If an appliance has a short circuit, it will draw excessive current, since the short presents a lower resistance path across the line than the regular circuit. This will blow a fuse or open a circuit-breaker. However, an open circuit-breaker or blown fuse may also result when too many appliances are connected on one circuit so that the total current drawn exceeds the rating of the circuit. Overloading is not usually dangerous, but it can be a nuisance.

1-2. Basic Principles of Appliance Repairs

Electrical appliances should give many years of trouble-free service. In the first place, they are relatively simple mechanisms that suffer little wear and tear with proper use. Secondly, they are used relatively infrequently, in comparison, say, to an automobile. For example, a car that is used only two hours a day (a moderate amount of use) will log more than 700 hours of operation in a year. A toaster, on the other hand, may be used only five or six minutes a day, or about 60 hours a year. It is not surprising that toasters can give twenty or more years of excellent service.

Most faults in electrical appliances are caused by misuse or outright abuse. A piece of silverware dropped in a garbage disposer doesn't help either the utensil or the appliance. An electric iron may fall off an ironing board. An appliance is disconnected from a wall receptacle by yanking on the cord. A coffee percolator is left on after it is empty. This sort of act of commission or omission may not immediately cause an appliance to fail, but it does shorten the useful life of the appliance. If an appliance is operated carefully in accordance with instructions, it almost never needs servicing.

Nevertheless, for one reason or another an appliance may need repair, and you have to decide whether to throw it out, send it to a repair shop, or try to fix it yourself. If the appliance is still covered by a guarantee, you have no problem. *Let the guarantor repair it! This is important, since you may void a guarantee if you take apart a faulty appliance, even if you do nothing else to it.* After the guarantee period has expired, you may decide to fix the appliance yourself.

Should you do this? If you like puzzles, you should be able to fix a faulty appliance and may even find it fun. First, you must determine what is wrong. Frequently, the trouble is in the plug or line cord and it is not necessary to take the appliance apart. You should be sure that the fault cannot be corrected externally before you try to disassemble the appliance. If you know that it is necessary to take apart the apparatus, you have to figure out how to do this. Manufacturers don't like people to take apart their appliances, figuring that these people will do more harm than good, both to the appliance and themselves. Therefore, many appliances are apparently put together in some mysterious manner, and solving this mystery is part of the problem. Some of the disassembly techniques are discussed in later chapters, but in general it is best to consider the appliance from every angle before removing a single screw.

An appliance may not operate properly, but as long as it gives adequate performance, it is not necessary to repair it. A toaster, for example, may toast bread unevenly, but as long as the toast is edible, its performance will probably be acceptable. As appliances age and their performance deteriorates, most users automatically "correct" for them. For example, if the temperature inside an oven becomes substantially different from that specified on the temperature control knob, a housewife usually becomes aware of the discrepancy and sets the control knob higher or lower as needed. Usually there is no complaint unless the appliance fails to work at all or blows a fuse when it is plugged in or turned on.

If an appliance doesn't work, that is, if nothing happens when the appliance is plugged in, it usually indicates an open circuit. This is shown in Figure 1-10 as a broken line cord. Note that when the line cord is plugged into the receptacle, the toaster will not work since the electrical circuit is open. In this figure, the break is shown as being in the line cord where it is easily accessible and can be repaired quickly. However, the open circuit could also be inside the toaster. It could be caused by a simple disconnection or by a burnt-out element. Remember: The fault may not be in the appliance at all, but in the wall receptacle. *This should be checked first*

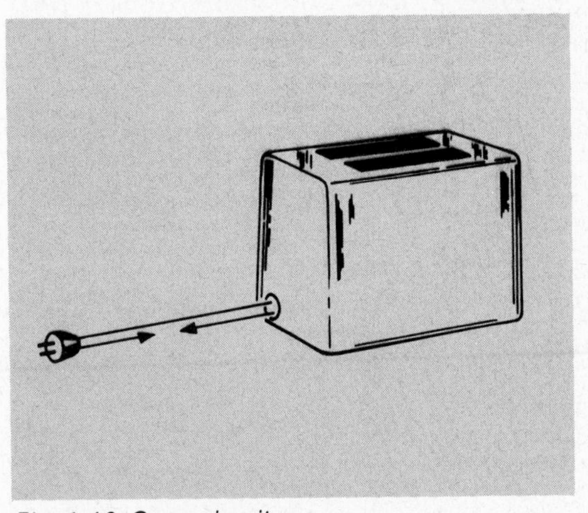

Fig. 1-10. Open circuit.

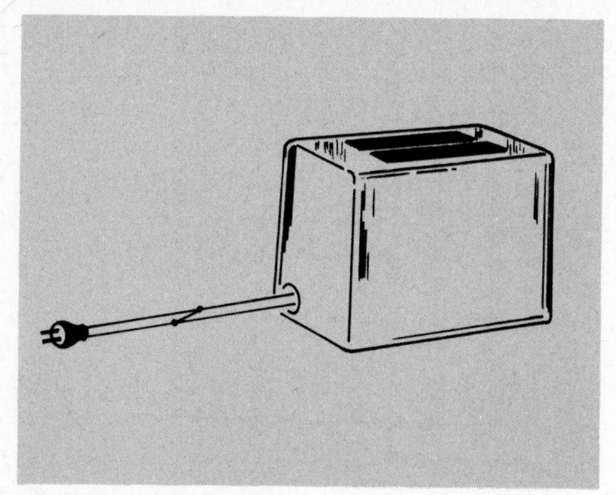

Fig. 1-11. Short circuit.

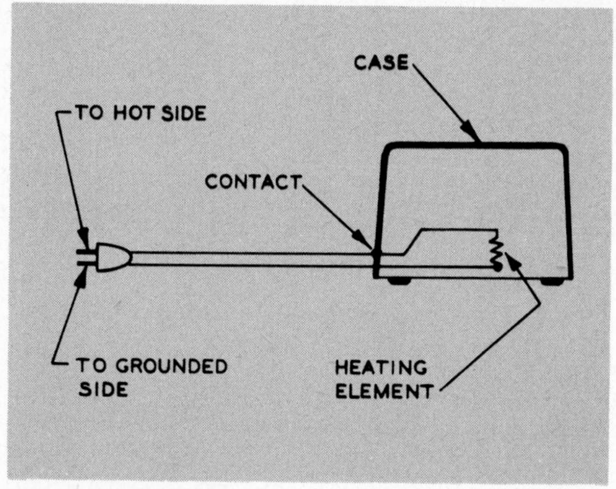

Fig. 1-12. Hot case.

whenever an appliance completely fails to operate. Plug in an appliance that you are sure is working.

If an appliance blows a fuse or trips a circuit breaker when it is plugged in, the trouble is usually a short circuit, as shown in Figure 1-11. Again, in the figure the trouble is indicated in the line cord. For example, the insulation between the two wires may break down from frequent flexing. The short circuit represents a path of very low resistance for electricity so that the current through it is very high. The excessive current trips the circuit breaker in the line. The short circuit could also be located inside the appliance, in which case it would be necessary to take the appliance apart to locate the fault.

It is also possible for an appliance to trip a circuit breaker if it *overloads* the line. This happens when several appliances are operated on one line simultaneously, drawing current close to the rated value for that line. Now if an additional appliance is connected, the current drawn exceeds the rated value, and the circuit breaker opens the circuit. However, this doesn't happen instantaneously as in the case of a short circuit; instead, the appliance works for a few seconds before the circuit breaker is tripped. The obvious solution when a line is overloaded is to move some of the appliances to another line.

Some problems are purely mechanical. For example, a toaster pops the bread entirely out of the slots when it is toasted, or a dishwasher leaks because of a faulty gasket. In each case, replacing the defective part solves the problem. Again, the major problem in repairing may be disassembling the appliance to get at the defective part.

A more dangerous flaw occurs in some appliances when the wire running to the hot side of the line somehow makes contact with the case of the appliance. This is shown in Figure 1-12. The defect can be caused by burnt or worn insulation or by a mechanical shock or vibration caused by dropping the appliance. It can sometimes be a defect in a new appliance which goes undetected for years, since the appliance still works perfectly even though the case is connected to the electric circuit. The potential danger of this flaw is that of electric shock, which can result if the case of the appliance and a cold water tap are touched simultaneously. A *temporary* solution is to pull out the plug, invert it and return it to the wall receptacle. Then the wire touching the case will be connected to the grounded side of the line, and there will be no danger of electric shock. However, this condition should not be allowed to remain for long, and the flaw should be corrected as soon as possible. Incidentally, every *new* appliance should be checked for this flaw before it is put in use, since it is something that can escape factory inspection, and it may be lethal. This check is described later in this chapter and again in Chapter 2.

1-3. Tools

Most electrical appliances can be repaired with ordinary hand tools. Although there are a few special tools designed to simplify repair of appliances, they are not worth buying unless you intend to go into the business. You should have an assortment of screwdrivers. Screws in electrical appliances come with a variety of different heads, as shown in Figure 1-13. Thus, you should have, besides the ordinary screwdrivers, a small Phillips screwdriver, an inexpensive set of Allen wrenches, and, desirable but not absolutely necessary, a set of nut-drivers which can also be used to tighten hex-head screws. A small crescent wrench is also handy if nut-drivers are not available. Wire strippers are useful, but an ordinary pocketknife serves almost as well. You will need a pair of long-nose pliers to get into tight spaces, and something for cutting wire, such as diagonal cutters or sidecutting pliers. For large appliances you may need a Stillson wrench, slip-joint pliers and a hammer.

When using a screwdriver on a small appliance or on a sheet metal surface of a large appliance, you must be careful not to strip the thread. Threads in sheet metal, plastic, or even in soft aluminum and brass are easily stripped by exerting too much force on the screwdriver. This danger can be minimized by using the *smallest* screwdriver that will do the job. Screws should be tightened using a light finger pressure, never by exerting torque from the wrist.

Since the electrical performance of an appliance must be checked before a diagnosis of fault can be made, some sort of diagnostic tool or instrument is required. Consider first the electrical checks you must make. You must find out if there is a short circuit or an open circuit in any part of the line cord or appliance. You also want to check to see that a voltage exists where it is needed and that there is no voltage where it may be dangerous (that is, on the case of the appliance). Note that accurate voltage readings or resistance values are not required. A volt-ohmmeter is helpful for making these measurements, but an inexpensive one will serve just as well as the most accurate, expensive models. If, however, you intend to service radios and TVs (not covered in this book), an accurate meter is necessary.

If you do not have a volt-ohmmeter and will repair only one or two appliances, you can make or buy a very inexpensive test-light which can be used to give you all the information you need to diagnose the trouble in a faulty appliance. It is simply an electric bulb in a socket with two wires connected to it, as shown in Figure 1-14. The wires are insulated, and pointed probes are connected to the free ends. Alternatively, if you wish to make a test-light yourself, you can use stiff

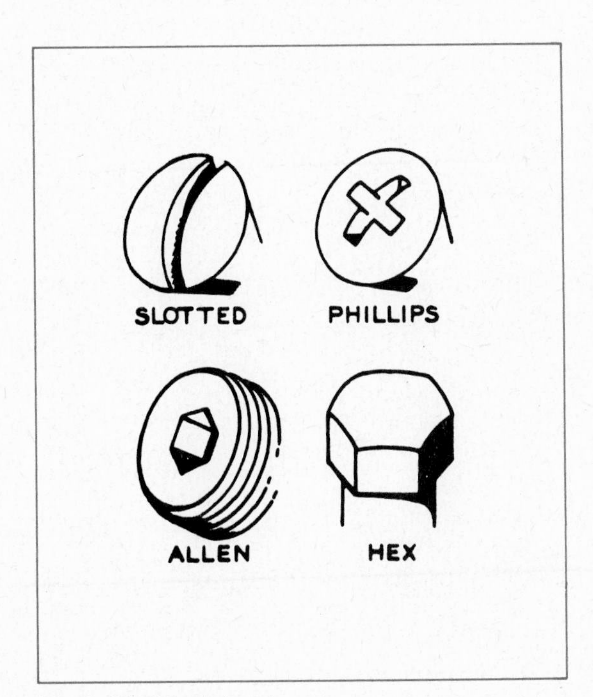

Fig. 1-13. Screw heads.

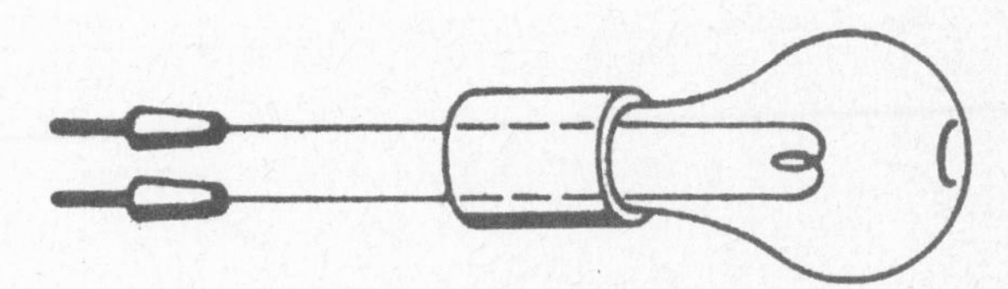

Fig. 1-14. Test-light.

wire and leave about one inch bare at the free end of each wire. The stiff wire protruding from the insulation makes a satisfactory test probe. Since flexible leads are easier to work with than stiff wire, you may want to use flexible wire, in which case you can stiffen the bare inch at the end by applying solder to it. You will also need a piece of wire two or three feet long with probes on both ends. One form of commercially available test-light uses a neon bulb instead of an incandescent bulb. It is very inexpensive and is a worthwhile addition to your tool kit.

Since the test-light described is used across the 115-volt line voltage, it must be handled carefully. Radio and auto supply stores also sell battery-operated test-lights which can be used without danger of electric shock. However, before spending too much money on a test-light, you should compare the cost with that of a cheap volt-ohmmeter.

A few clip-leads will simplify servicing. These are short pieces of wire, about six to eighteen inches in length, with alligator clips at both ends. By clipping the alligator clips to various points in the circuit, it is possible to tie parts together electrically or to bypass parts of the circuit so that other parts may be checked alone.

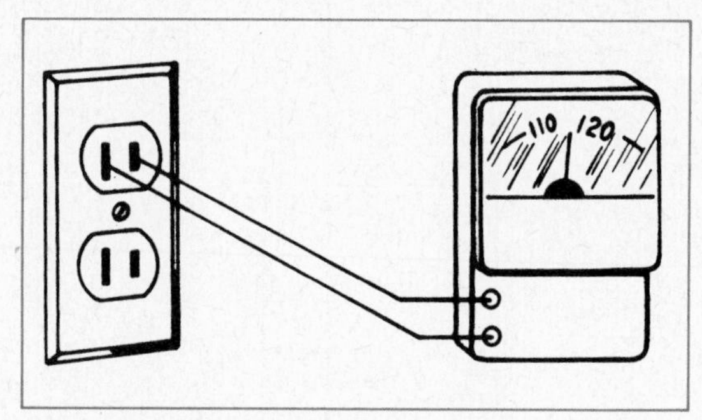

Fig. 1-15. Voltage check with meter.

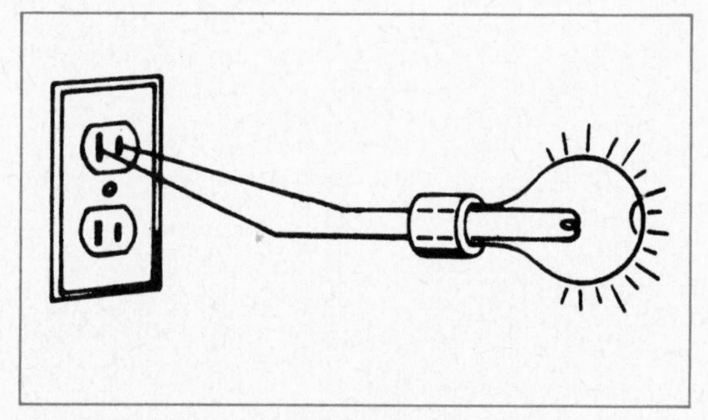

Fig. 1-16. Voltage check with test-light.

1-4. Basic Tests

The primary test is a voltage check to make sure there is voltage where it is needed and no voltage where it may be dangerous. This check can be performed with a test-light as well as with a volt-ohmmeter, and indeed, for the vast majority of troubles the test-light is adequate. Another important test is a continuity check to make sure the circuit is continuous. A test for a short circuit is a type of continuity check where you hope the circuit will not be continuous. A continuity check is performed easily using the resistance scales of a volt-ohmmeter. It can be done with a test-light but that requires more care.

When an appliance doesn't work at all, the first thing to determine is whether there is voltage at the wall receptacle. This can be ascertained easily by plugging in a different appliance which is known to be in working condition or by checking the outlet with your volt-ohmmeter or test-light. If you use a meter, make sure that you set it for a scale higher than the expected voltage. Therefore, if you are measuring a 115-volt line voltage, use the 150-volt or 250-volt scales. If you use a low scale, for example, the 10-volt scale, the 115 volts will knock the pin off the scale and may damage the meter. The two leads from the meter are inserted in the receptacle as shown in Figure 1-15, and the meter should indicate a voltage between 110 and 120 volts. The exact reading doesn't matter. Similarly, if you use a test-light, the two leads are inserted into the receptacle in the same manner, as shown in Figure 1-16. If the bulb lights, there is voltage at the receptacle. When inserting leads into a receptacle, make sure you hold them by their insulation, and never touch the bare wire! If it is difficult to do this, it may be easier to use one side of each half of the receptacle, as shown in Figure 1-17. The bulb

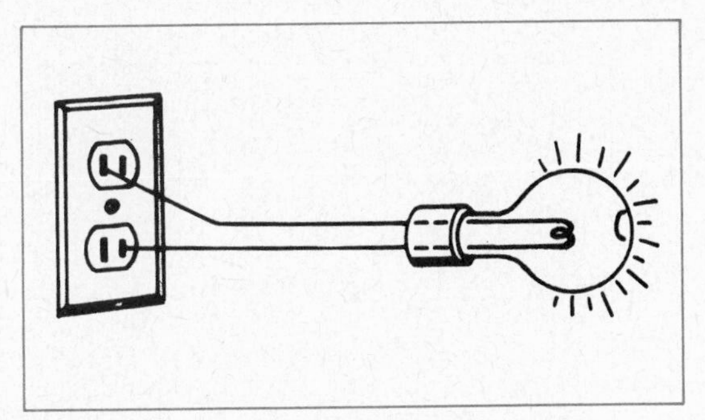

Fig. 1-17. Alternative method of checking receptacle.

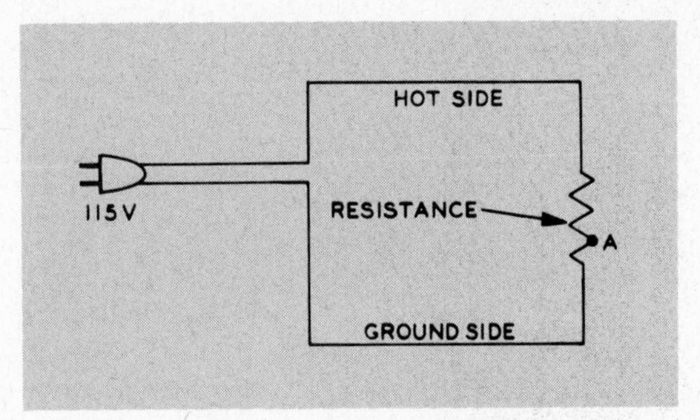

Fig. 1-19. Voltage drop across resistance.

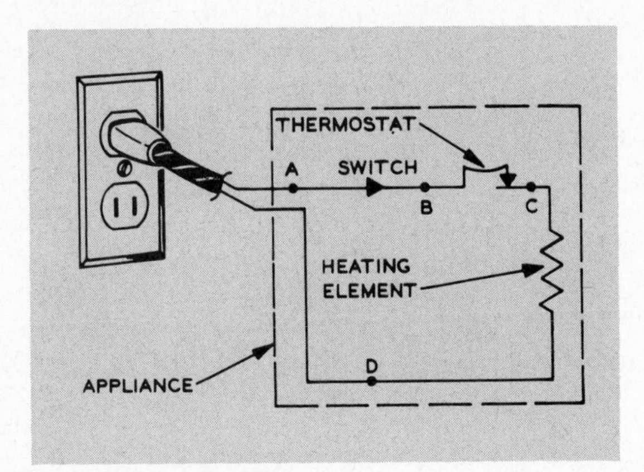

Fig. 1-18. Test points.

will light as before, since both halves of the receptacle are wired in the same way. If the bulb fails to light, the trouble may be in the receptacle or it may be a tripped circuit breaker. Before tackling an appliance itself, make sure the receptacle is "hot".

Assuming the receptacle is hot, the appliance is now plugged in, but nothing happens. Voltage readings can be made at several points inside the appliance using the test-light or the meter. In Figure 1-18 several test points are indicated. If the two wires from the tester are connected to points A and D in the figure, a voltage reading would indicate that the cord is satisfactory. If now the top probe is moved to point B, holding the bottom one on D, failure to light the bulb or get a meter reading will indicate a faulty switch, which should then be replaced. On the other hand, if the switch is all right the next check might be between points C and D. A lack of voltage here would indicate a bad thermostat. Note that a thermostat is simply a switch operated by heat. If the thermostat checks out all right then the trouble could be in the heating element. In practice, a visual inspection might reveal the open circuit without the necessity of all the voltage checks, but the purpose here is to show that voltage can be measured anywhere in the circuit.

When a resistance is connected across the 115-volt line, there are 115 volts across the resistance, and this voltage causes current to flow through the resistance. If one end of the resistance is connected to ground, as shown in Figure 1-19, the hot side of the resistance is 115 volts different from ground. There is said to be a *voltage drop* across the resistance. If we were to connect one lead from our meter to a fixed ground connection and the other lead to the hot side of the resistance, we would read 115 volts on the meter. But if the second lead were connected to the ground side of the resistance instead of the hot side, we would read zero volts. If the lead were slid along the resistance from the hot side toward the ground side, keeping the other lead fastened to ground, the meter reading would begin at 115 volts and decrease gradually to zero as the lead was moved along the resistance. For example, in Figure 1-19, if the hot lead to the meter contacted point A of the resistance and the other lead was grounded, the meter would read about 50 volts, since A is somewhat nearer the ground side than the hot side of the resistance.

The discussion in the preceding paragraph indicates the need for a voltage check on appliances. Assume the resistance in Figure 1-19 is a heating element in a toaster. Assume also that because of carelessness in shipping, point A of the heating element has somehow made contact with the case. The toaster will still toast bread perfectly, but the case now has a voltage of 50 volts on it. If someone touched the case and a cold water faucet simultaneously, he would get a shock. This is a hazard in every appliance, and consequently every new appliance and every repaired appliance should be checked for this danger. To check, simply touch one lead from the meter to the water faucet and the other to the case of the appliance. The voltage should read zero. Notice that even if the voltage is only five volts, it indicates that something is wrong. Five volts will not give you a shock or even a tingle, but if someone pulls out the plug and then puts it back inverted, the ground side and the hot side will be interchanged. This means that the case, which before was five volts away from ground is now five volts away from 115 volts. The resultant 110 volts is dangerous.

You can also use a test-light to make this check. Place one lead on the water faucet and one on the case. If there is 50 or 60 volts on the case, the light will glow at about half its normal brilliance. If there is substantially less, there might still be a hazard, but not enough voltage to light the bulb. However, if the plug is removed from the receptacle and inverted, there will be sufficient voltage to light the test-light. Thus, whenever you use the test-light, you should keep this possibility in mind and make checks with the plug both right side up and inverted.

If two points are connected by a short circuit, there should be zero voltage difference between them. A switch is supposed to be a short circuit when it is operating properly. However, every time a switch is opened or closed, a small electric arc jumps between the connecting points and tends to burn these contacts. After many switchings the contacts may be so burned or dirty that the switch is no longer a good circuit, but presents some resistance to the current. There is then a voltage drop across the switch and thus, less voltage across the appliance itself. The appliance might still operate but, for example, a toaster may take longer to toast bread, or a vacuum cleaner may have less suction. To check a switch, the meter leads are placed across it. Referring to Figure 1-18, one lead from the meter is placed in contact with point A and the other with point B. With the switch open, you should read 115 volts, since point A is connected directly to one side of the house line through the line cord, and point B is connected to the other side through the appliance itself. When the switch is closed, the voltage should be zero. The meter scales should be changed downward, with the leads contacting points A and B, to increase the sensitivity. If no reading is obtained on a 5-volt or lower scale, the switch is probably satisfactory. Important: *Always return the scale setting to something more than 120 volts before making further checks.*

A test-light like that illustrated in Figure 1-14 cannot be used in this manner to check a switch, since the voltage across the switch may be insufficient to light the bulb. However, the inexpensive neon tester is sensitive enough to detect a very small voltage and can be used to check a switch.

Another type of test is a continuity check of the circuit or parts of the circuit. By means of this test you determine if the circuit is continuous or whether it has a break in it. The ohmmeter part of the volt-ohmmeter is used. The appliance to be tested is disconnected from the house line. Usually when you perform this test you want to know if the circuit is complete and you are not interested in exact values of resistance. The two leads from the ohmmeter are connected to two points in the circuit. If there is a break in the circuit, the needle on the meter will not move, but if there is continuity, the needle will swing to the right. When there is very little resistance in the circuit, the needle swings full scale to the right, indicating a short circuit; otherwise, it gives some indication of the amount of resistance between the points contacted. These points can be anywhere in the circuit, as shown in Figure 1-20. To start, the two leads may be connected to the prongs of the plug, points A and B in the figure. If the needle swings toward the right, there is a continuous circuit. However, this does not necessarily mean that the circuit in the appliance is continuous, since a short circuit in the line

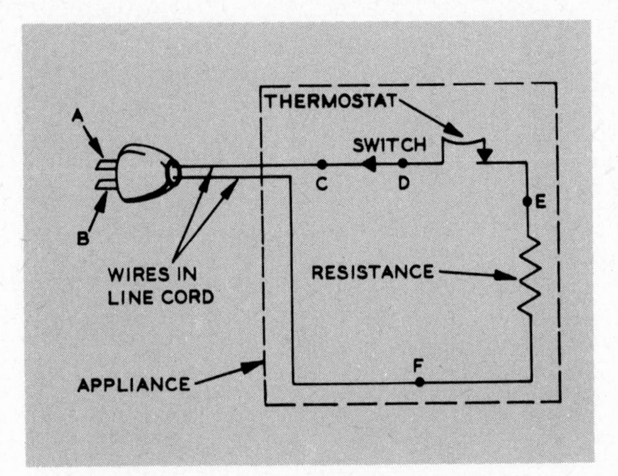

Fig. 1-20. Test points for continuity check.

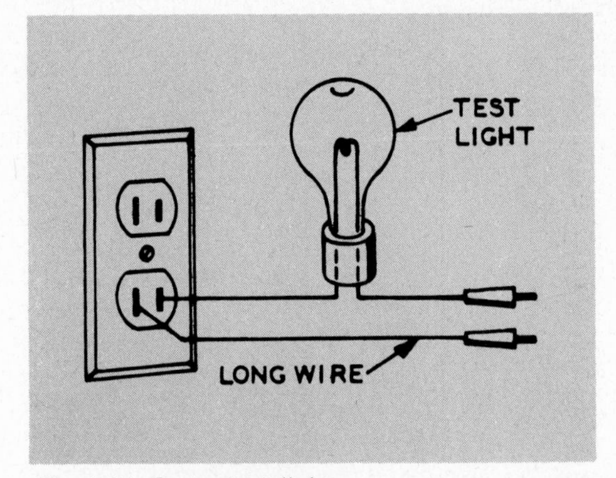

Fig. 1-21. Series test-light.

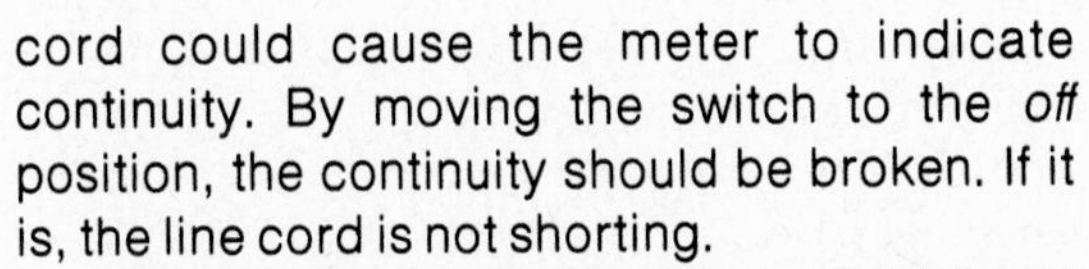

cord could cause the meter to indicate continuity. By moving the switch to the *off* position, the continuity should be broken. If it is, the line cord is not shorting.

Checks can be made between points C and D to determine if the switch is working. Here the indication should be a short circuit when the switch is closed and an open circuit when it is open. The resistance in the circuit in Figure 1-20 may be a heating element, a motor, or any other electrical component. By checking between points E and F you can determine if there is a break in the wiring in this part of the circuit.

It is possible to use a test-light to make a continuity test but it requires a lot of care since it involves the line voltage as well. One lead from the test-light is plugged into one slot of a receptacle and a long wire (2 or 3 feet) is plugged into the other. The setup is shown in Figure 1-21. This arrangement is sometimes called a *series test-light.* Now if the two probes at the right are brought together, the bulb will light, since the circuit will be complete. Again, if you touch these probes to points C and D in Figure 1-20, the bulb will light if the switch is closed, but will stay off if the switch is open. The series test-light can be used in the same manner and same places as the ohmmeter. Continuity is indicated by the lighting of the bulb.

Frequently when an appliance fails, the trouble will be obvious. For example, a switch fails to "click" when pushed to the *off* position, or a wire inside the appliance is obviously broken. You can make a quick check to determine whether the apparent trouble is in fact the actual trouble. For example, you suspect the switch. *With the appliance disconnected*, take a short clip lead and connect it across the switch. That is, clip one end to point C and the other to point D in Figure 1-20. The clip lead will complete the circuit if the switch failed to do so. Now plug in the appliance. If it works, the switch must be at fault and should be replaced. In the case of the broken wire, use a clip lead to fasten the two ends together. Now when you plug in the appliance and turn it on, it should work if the only trouble was the broken wire.

2 Safety

2-1. Electricity Can Be Lethal

The fact that electricity can kill you is something you must keep in mind whenever you work on any kind of electrical appliance. You might be lucky and escape injury if you carelessly touch a live wire. You might also avoid an accident if you drive an automobile the wrong way on a one-way street. Common sense dictates that both actions should be avoided. Many of the warnings in this chapter also appear in the first chapter, but because of the danger involved, they bear repeating.

Always approach the repair of an electrical appliance with caution. This means giving the task your undivided attention and thinking constantly of the danger. A shock caused by carelessness is just as dangerous as one due to ignorance. You must also bear in mind that after the appliance is repaired, it must be safe to operate. That is, there must be no danger of shock to the person who uses the appliance in a normal manner. There is nothing you can do about someone who uses a metal knife to pry out a piece of bread stuck in a toaster while the toaster is on, but you *can* make sure that he won't get a shock if he touches the case of the toaster and the water faucet at the same time.

There are three simple rules of safety when you are working on appliances. If they are followed *at all times*, you will avoid danger to yourself and to the user.

1. *Do not touch any bare wire* which is connected to an electrical outlet.

2. *Never work on anything "live".*

3. *Always* check to make sure that there is no voltage between the case of the appliance and ground.

The first rule is fairly obvious and is usually carefully observed by beginners, but it is sometimes disregarded by sophisticated repairmen who should know better. It should never be disregarded. *Never* touch a bare wire which is connected to a source of voltage even if you *know* that it is safe. Consider the simulated appliance shown in Figure 2-1.

The line cord is plugged into a receptacle, but the switch in the appliance is in the *open* position. Point A to the left of the switch is now connected directly to the hot side of the line. Thus, point A has a potential of 115 volts from ground. Point A is a dangerous point to touch, especially if you may also be touching ground. (If you are standing on a concrete floor in leather-soled shoes, you are touching ground.) However, you know that point C is connected to the ground side of the line and should be safe to touch. Also, point B is

When working on an electrical appliance, always make sure the plug is in plain view.

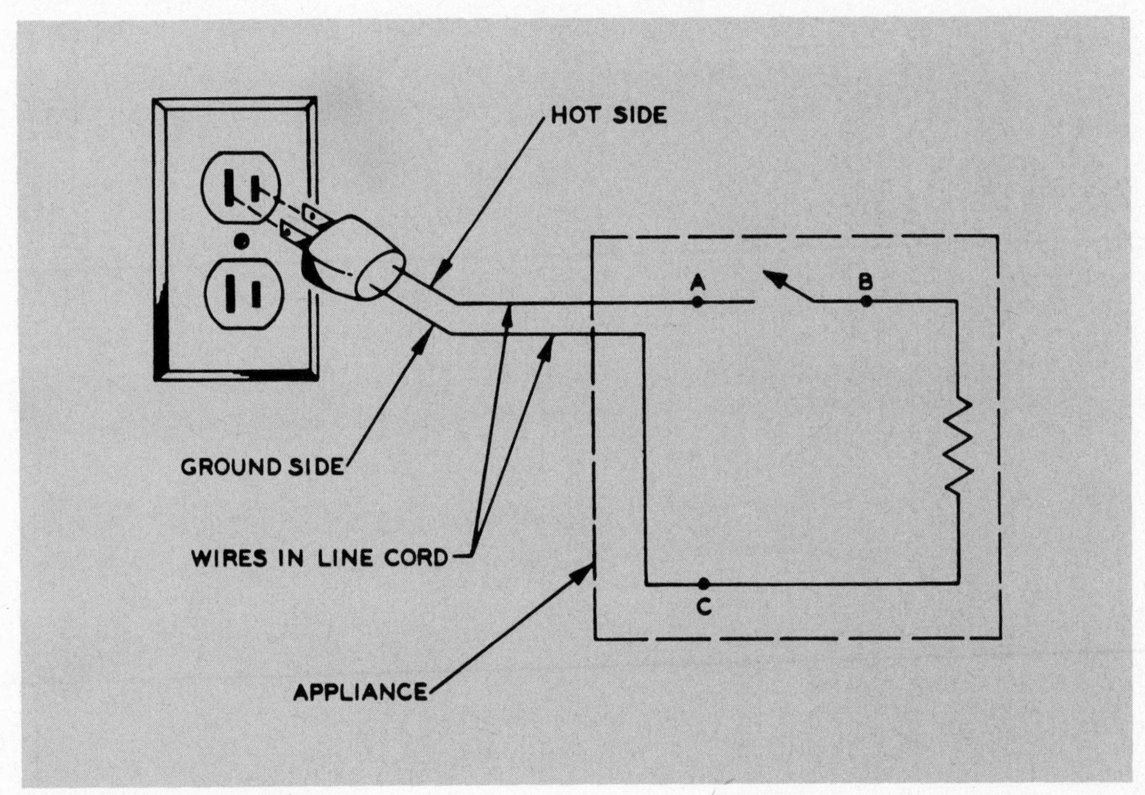

Fig. 2-1. Simulated appliance.

connected to point C through the appliance and therefore should also be safe to touch. The trouble is that you know too much. It is true that you can safely touch points B and C in Figure 2-1. This, however, is a bad habit to get into. In the first place, you might make a mistake as to which is the hot side of the line. Secondly, you might pull out the plug and then put it back upside down so that point A is now grounded, and points B and C are hot. Worse, someone else might do this while your attention is elsewhere. Therefore, never touch a bare wire unless it is disconnected from all voltage sources.

When you are making voltage checks with a meter or a test-light, it is necessary to hold the probes in contact with points of high voltage in the wall outlet or in the appliance being tested. Always hold the probes by the insulation. Again, never touch a bare wire or a bare probe.

The second rule, never work on a "live" appliance, refers to actual repairs. Of course, you must plug the appliance in and turn it on in order to make voltage checks on it. In this sense, you are working on a live appliance. However, once you have located the trouble, you must pull out the plug before making a single change or removing a single screw. Note that it is not enough to shut off the switch. If the appliance is plugged in, there are points of high voltage, (for example, point A in Figure 2-1). Make it a habit to pull out the plug and *have it lying in plain view* while you are working on the appliance. In this way you will not get a shock because you *thought* you pulled the plug.

The third test, the ground test, checks the appliance for safety to the user. The purpose of this check is to make sure that the case of the appliance does not have a voltage on it. Before you do this, it is well to check your outlets to determine which side of the line is grounded. If the receptacle has been wired correctly, the longer slot is the grounded side of the line, as shown in Figure 1-4. However, sometimes a home owner decides to install a new receptacle himself and does so incorrectly. Then the short slot becomes the grounded side. Using a test-light or meter, put one probe in a slot in the receptacle and touch

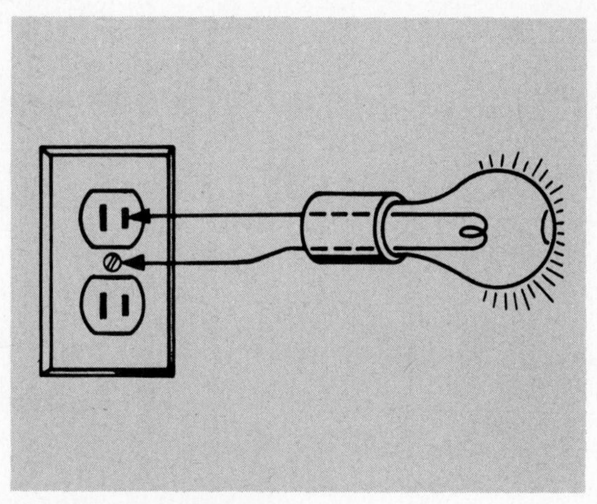

Fig. 2-2. Ground screw.

the other probe to a good ground connection, such as a water pipe. If the bulb lights or the meter reads about 115 volts, the probe in the receptacle is in the hot side of the line. Check the other side, anyway. The light will not go on if you have the wall probe in the slot connected to the grounded side of the line.

If a wall outlet is installed correctly, the mounting screw in the center is connected to ground through the shield that encloses the wires in the wall. Thus, the screw can be used as a ground connection when a water pipe is not available. This is illustrated in Figure 2-2. One probe can be placed on the screw and the other in the hot slot, and the bulb will light. You must be sure that the screw is not covered with paint or other insulating material.

To make the ground safety test, plug in your appliance and check to see if you have voltage between the case and a good ground. If you have ascertained that the screw on the wall outlet is a good ground, you can use that; otherwise, use a water pipe or faucet. The test is illustrated in Figure 2-3. A toaster is shown in the figure, but the same test could be used for any appliance. One probe of the meter is held in contact with the case of the appliance and the other probe is connected to the good ground. In the figure, the screw in the center of the receptacle is used as a ground.

If there is any voltage reading at all, even as little as one volt, there is a potential danger. A low voltage won't give you a shock, but it does

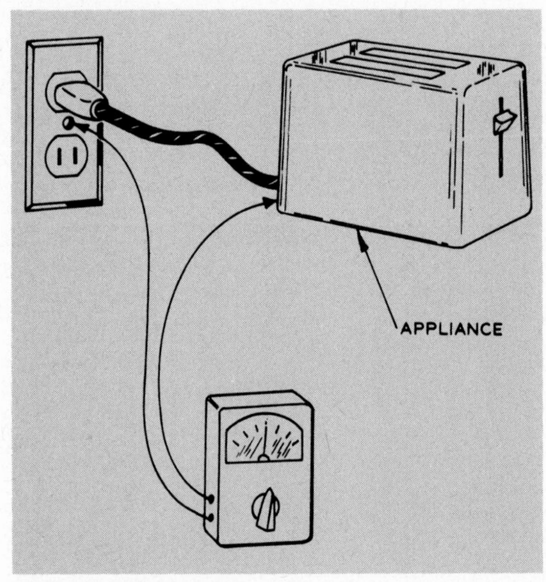

Fig. 2-3. Ground test.

indicate a faulty connection or weak insulation, and it can only get worse. The trouble should be corrected before it becomes dangerous. Note that a normal test-light may not be sensitive enough to detect a low voltage, but a neon tester will. If you do not have a meter, a neon tester is a necessary tool along with a test-light.

In making the ground test, make sure that you pull out the plug and invert it so that the grounded wire in the line cord becomes the hot wire, and vice versa. Make the test for both positions of the plug with the appliance switched on and with it off. There should be no voltage between case and ground in any of these tests. If there is a voltage in even one position of the plug and switch, the appliance is not ready for use until the cause of the trouble is eliminated.

Large appliances such as clothes dryers should be grounded to a water pipe.

One way of making certain that the case of an appliance will never be hot is to permanently fasten the case directly to ground. This should be done on all large appliances that are never moved, such as an oven, clothes washer or clothes dryer. If you have a wall receptacle with a ground terminal as shown in Figure 1-5, then it is a simple matter to run a three-wire line cord to the appliance. The third wire is fastened to the case at the appliance end and to the round prong on the plug. Even without this type of grounding outlet, you can still ground the appliance by running a wire from the case to a ground connection. Use a piece of wire which is rated to carry the current drawn by the appliance. Loosen a screw on the case (preferably in back and out of the way), wrap an end of the wire around the screw, and tighten the screw again. Connect the free end of the wire to a grounding clamp which is attached to a water pipe. Grounding clamps are available in hardware or electrical supply stores. The case of the appliance will then be firmly connected to ground. If subsequently something shorts inside the appliance, putting a voltage on the case, it may cause a fuse to blow, but it will not endanger the life of someone touching the case.

Repairing electrical appliances must not be done haphazardly. Neatness is important, not only because it looks good, but because it is safer. For example, a clean connection has no stray wire sticking out which may contact the case of the appliance or touch another wire causing a short circuit. Connections should be "solid" so that the applaince will stand the abuse that careless users too frequently give them. Although everyone knows that a line cord should not be yanked out of an outlet, people do so frequently. Consequently, wiring must be done in such a way that this type of abuse will put no strain on the electrical connection. It is also obvious that wires must not interfere with moving parts in appliances.

3 Line Cords

The most common source of trouble in an electrical appliance is the line cord and its connectors. Although the repair or replacement of line cords and the installation of new connectors are relatively simple tasks, more often than not they are repaired improperly, leading to further trouble. It is important to know something about wires and connectors to be sure that you select the proper material for the job.

3-1. Cords

Since line cords must be flexible to go around objects or to permit movement (with vacuum cleaners and electric irons, for example), they are always made of stranded wire. If a single solid wire were used for each conductor, it would have to be large enough to carry the current required by the appliance. For most

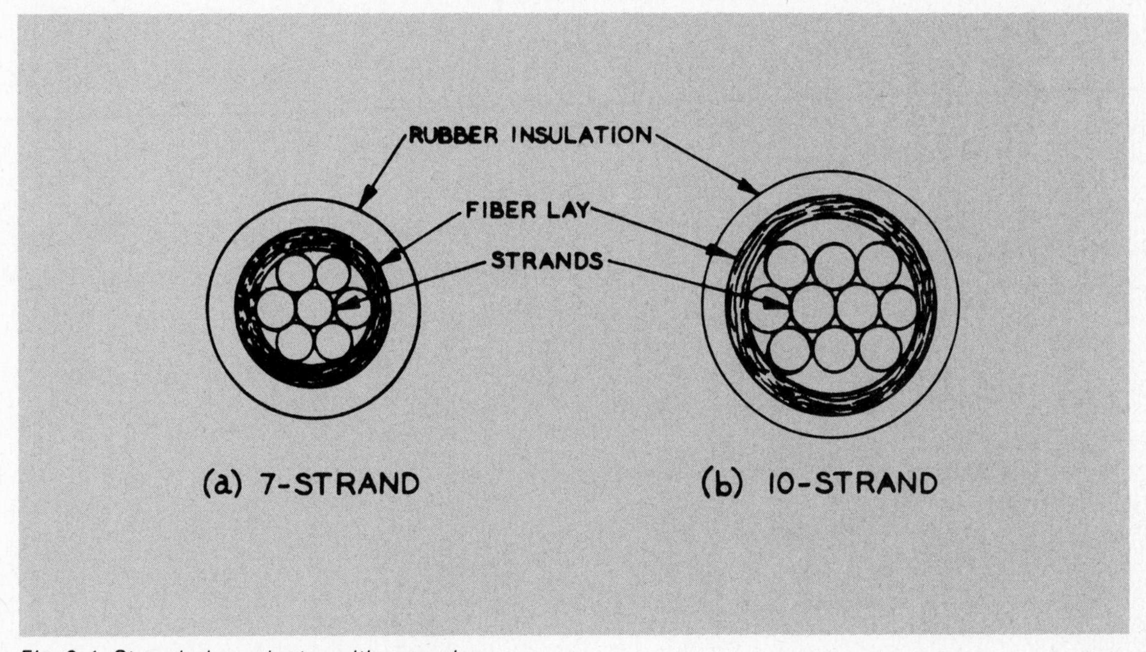

Fig. 3-1. Stranded conductor with wrapping.

appliances, the wire size required could be achieved only in a stiff wire. Thinner wire is flexible, but will not carry sufficient current without overheating. However, if several thin wires (or strands) are grouped together in a single conductor, they share the current equally, and thus can carry heavier currents without overheating. All line cords use stranded conductors. Cross sections of individual conductors are shown in Figure 3-1. Each conductor consists of a bundle of strands of thin wire. This bundle is tightly enclosed with a wrapping of fiber, called a *lay*, and then covered with rubber insulation. There are several different types of line cords, but all those used with home appliances have the conductors wrapped in fiber and rubber, as shown in Figure 3-1. The more popular line cords have conductors with seven or ten strands, but heavy-duty wires with as many as 65 strands in each conductor are also used.

When you strip insulation from a conductor in a line cord, it is important that you do not cut into any of the strands. If you break a few strands, the appliance will operate properly, but each of the remaining strands will carry more current than it should. For example, suppose a toaster which draws 10 amperes has a line cord with 10-strand conductors, as shown in Figure 3-1 (b). Each strand then carries only one ampere, and these strands can be relatively thin wires. However, if six strands break because of too much flexing or poor installation, the remaining four will have to carry the 10-ampere load, and each will carry 2.5 amperes. This will cause the line cord to get hot in the vicinity of the break. You can use a jack-knife to strip insulation, but you should not use too much force. Cut into the insulation *lightly* by rotating the knife around the wire about an inch from the end, as shown in Figure 3-2. When the insulation is cut all the way around, the severed portion can be pulled off like a sleeve. The fiber lay between the strands and the rubber insulation protects the strands from the knife, and there should be little risk of cutting the wire strands as long as the job is done carefully. After pulling off the sleeve, the fiber lay can be unwrapped, as shown on the conductor at the left in Figure 3-2, and can be cut off with a pair of scissors or a knife.

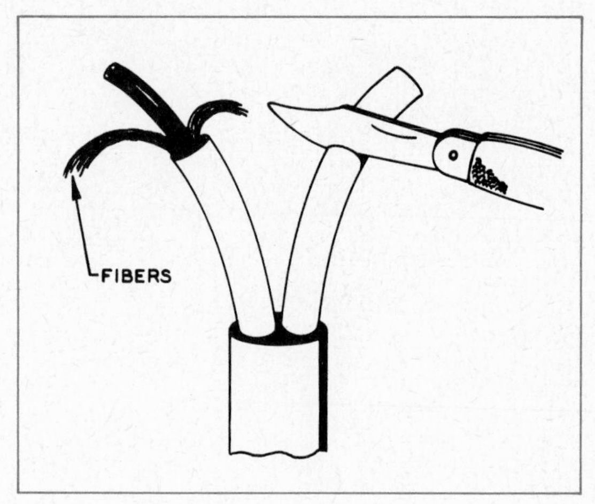

Fig. 3-2. Stripping insulator.

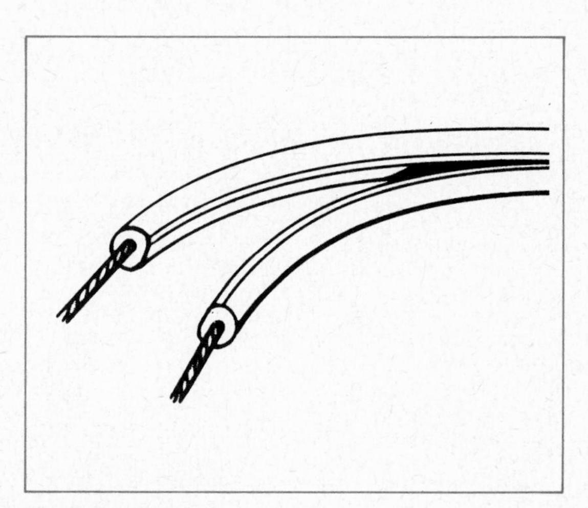

Fig. 3-3. Zipcord.

For most low-power appliances, such as clocks, juicers and lamps which draw less than two or three amperes, the most common type of line cord is the popular *zipcord*, shown in Figure 3-3. This consists of two conductors, like those in Figure 3-1, joined together by a thin rubber bond. It is called zipcord because the two conductors are easily separated by tearing the thin rubber membrane joining them.

For heavier work, such as mixers and vacuum cleaners, *jacketed cable* is used. The

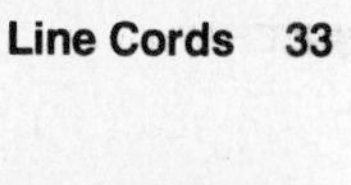

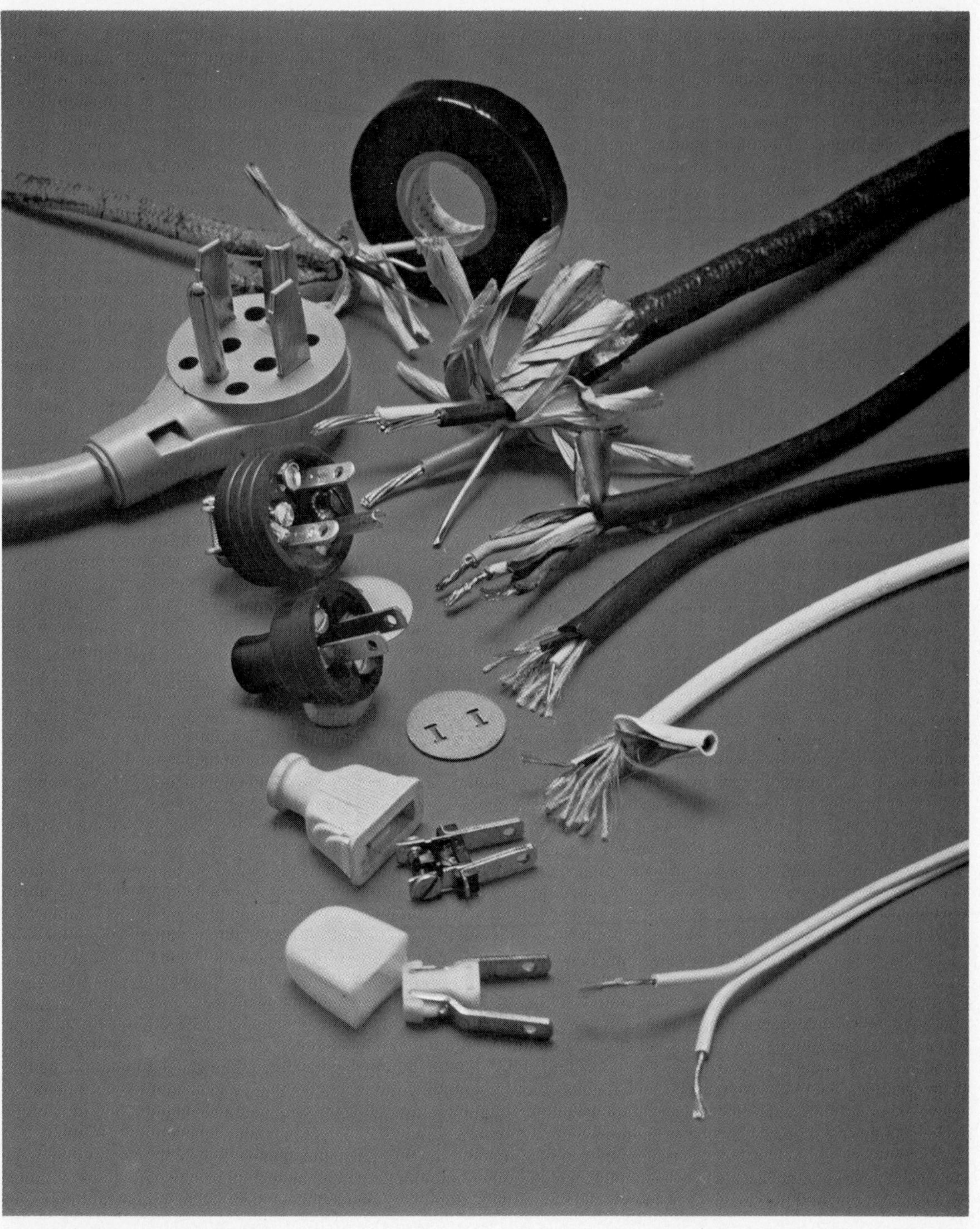

Types of line cords and plugs.

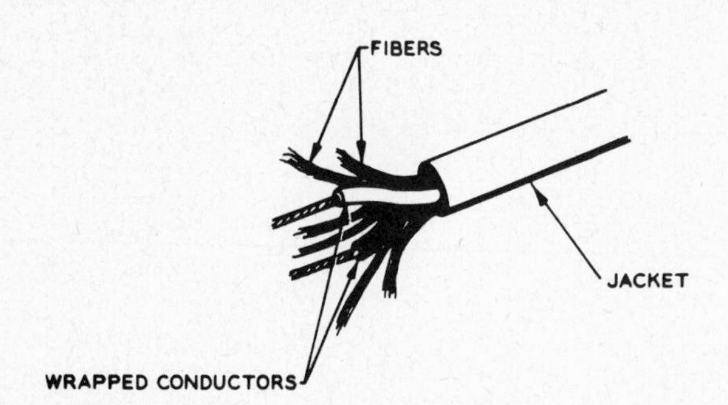

Fig. 3-4. SV cord.

wrapped conductors are separated by additional fibers which increase the insulation between them, both electrical and thermal. The two conductors and extra fibers are encased in a rubber or plastic jacket. This type of line cord is called "SV cord", and its construction is shown in Figure 3-4. In general, jacketed cable will take more abuse than zipcord.

For appliances which draw more than five amperes, special *heater* cord is preferred. This is used on toasters, waffle irons, irons and other appliances furnishing heat rather than motion. In heater cord, each wrapped conductor shown in Figure 3-1 is further wrapped with asbestos fibers. The two conductors are then brought together and a layer of asbestos is wrapped around both. Finally

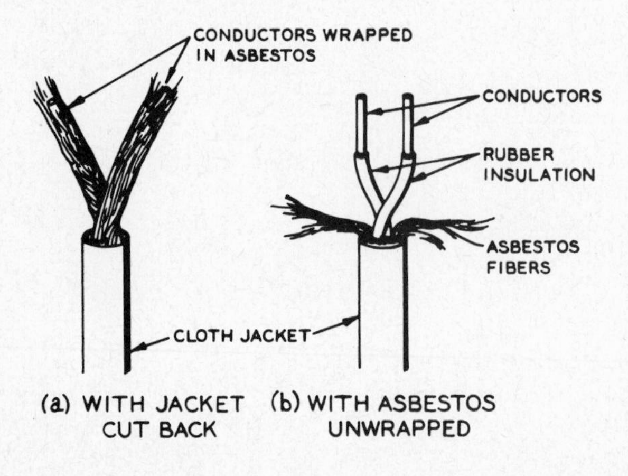

Fig. 3-5. Heater cord.

the package is encased in a braided jacket of cotton or nylon. Heater cord is illustrated in Figure 3-5.

The current-carrying capacity of line cord depends on the wire diameter and not on the insulation. Thus, a zipcord line with # 10 wire can carry 25 amperes, whereas an asbestos heater cord with # 18 wire may be limited to 5 or 6 amperes. In general, heater cords are used for appliances that draw more current; jacketed cable is used where the cord itself is subject to abuse; and zipcord use is by far the most common because of its simplicity. Zipcord is used wherever there are no special demands on the line cord.

3-2. Connectors

Most of the early appliances did not have line cords connected to them, but instead a *cord set* was supplied with each appliance as a separate unit. A cord set consists simply of a line cord with a connector at each end, one a wall plug to connect to the wall outlet, and the other an appliance plug to connect to the appliance. Some appliances, such as irons and vacuum cleaners, are moved frequently in use, and this motion causes excessive bending and strain at the appliance end of the cord, which often causes the wire to break inside the connector or to become disconnected. To eliminate this source of failure, manufacturers now tend to have the appliance cord fastened to the appliance internally, although a few appliances still come with separate cord sets. Another advantage of an integral line cord is that a user cannot connect the wrong cord set to an appliance.

Wall plugs come in a large assortment of shapes, materials and colors. A few types are shown in Figure 3-6. The molded plastic plug in (a) is usually furnished with new appliances. The connections between wires and prongs are tightly sealed, and with proper use this type of plug should outlast the appliance.

The shell-type plug shown in (b) is a

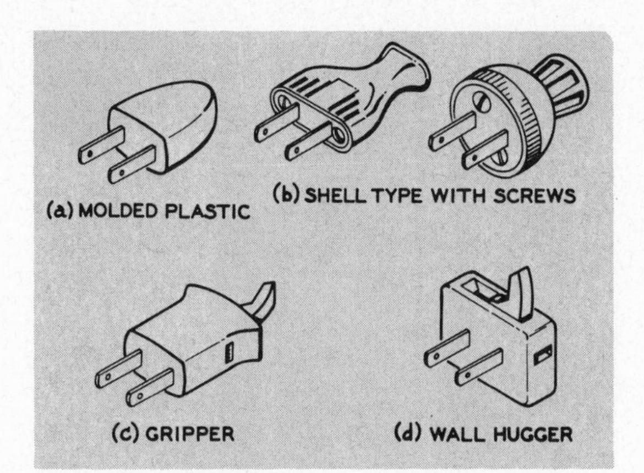

Fig. 3-6. Plugs.

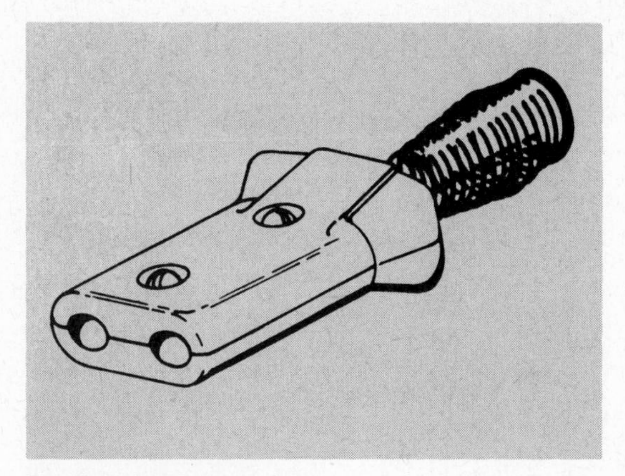

Fig. 3-7. Appliance plug.

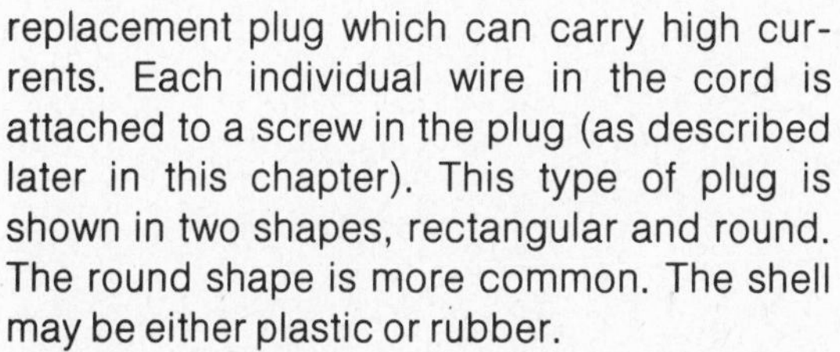

replacement plug which can carry high currents. Each individual wire in the cord is attached to a screw in the plug (as described later in this chapter). This type of plug is shown in two shapes, rectangular and round. The round shape is more common. The shell may be either plastic or rubber.

The gripper-type plug shown in (c) is the simplest to install, but is suitable for use only on zipcord for low-current applications. To attach this type of plug, you simply cut the zipcord squarely, and without stripping off any insulation insert it in the hole in the side of the plug as far as it will go. Then push down the lever on the back of the plug, which causes small teeth on the connectors to bite through the insulation and make contact with the wires. The *wall-hugger* plug shown in (d) is the same kind of connector, but has the prongs protruding from the largest face, so that when it is inserted into a wall receptacle, the plug extends less than half an inch from the wall and causes a minimum of interference with furniture next to the wall. Wall-hugger plugs are also available with screw connections for handling higher currents.

An appliance plug is shown in Figure 3-7. This plug comes in two halves which are held together with either screws and nuts or rivets. Inside are two heavy-duty contacts which slide over projecting pins in the appliance. These contacts have screws to which the wires in the cord are connected. The spring at the back of this plug prevents sharp bends in the cord which may damage the insulation or the wires.

3-3. Faults and Repairs

Many old line cords on appliances were covered with reclaimed rubber which does not stand up well over a long period of time. If these cords are exposed to sunlight for long periods or stretched behind radiators, the rubber becomes dry and brittle, and sometimes pieces break off in chunks, leaving exposed wire. Needless to say, this dangerous condition must be corrected as soon as it is detected. If a visual inspection of any cord shows cracks or dryness in the insulation, you should flex the cord sharply to see if the insulation can stand it. If not, replace the cord.

Modern line cords should outlive the appliances they are attached to. However, modern users of line cords tend to abuse the cords and cause problems. The most common fault is pulling a plug from the wall outlet by yanking the cord. There are two dangers here. You may pull the wire from its screw contact, if it was improperly connected, or, secondly, you might break some or all of the strands in a wire. Another problem arises when something is pushed against a plug in a wall receptacle so that the cord entering the plug is bent

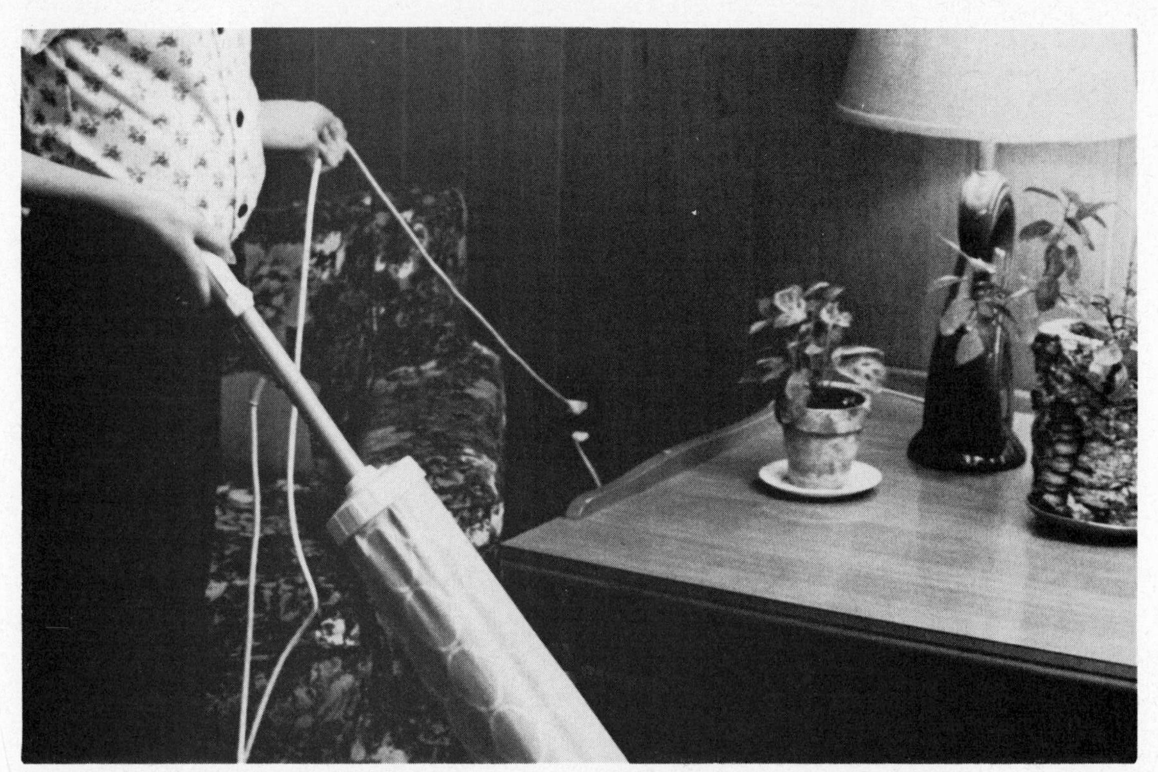

Never yank a cord from the outlet as this will eventually damage the plug and the cord, and create a hazardous condition.

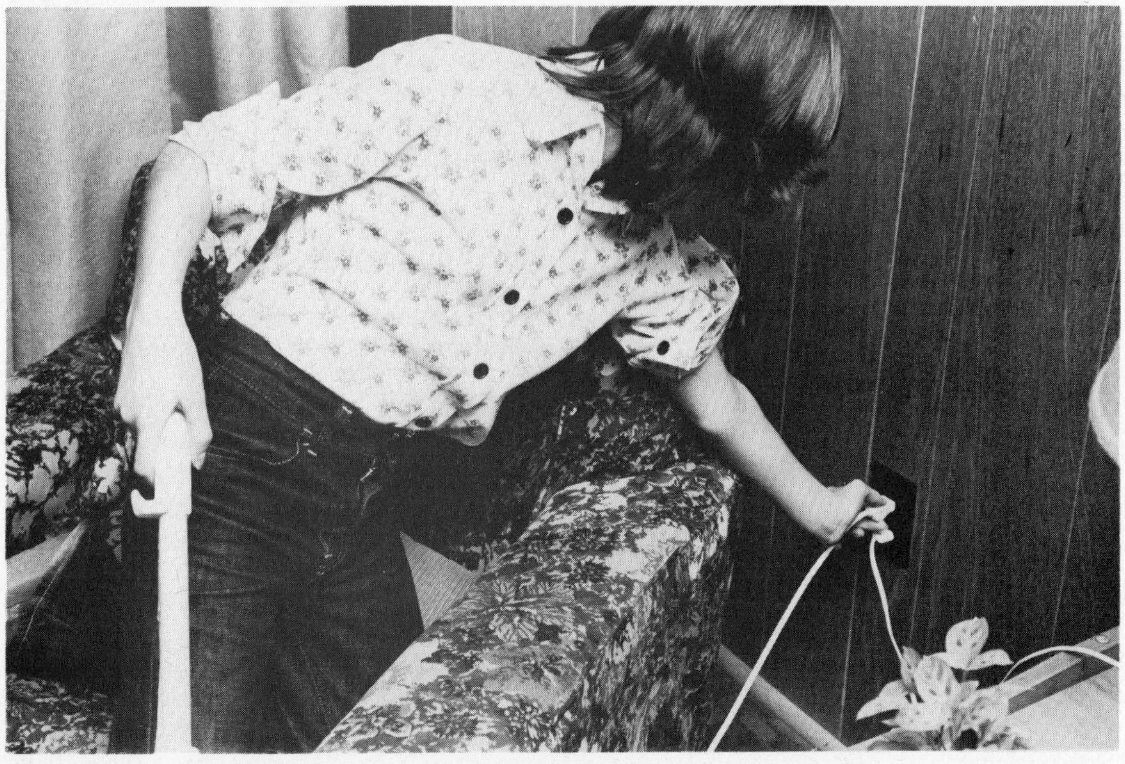

When pulling a plug, always take hold of the plug itself; don't yank the cord.

Vaporizer.

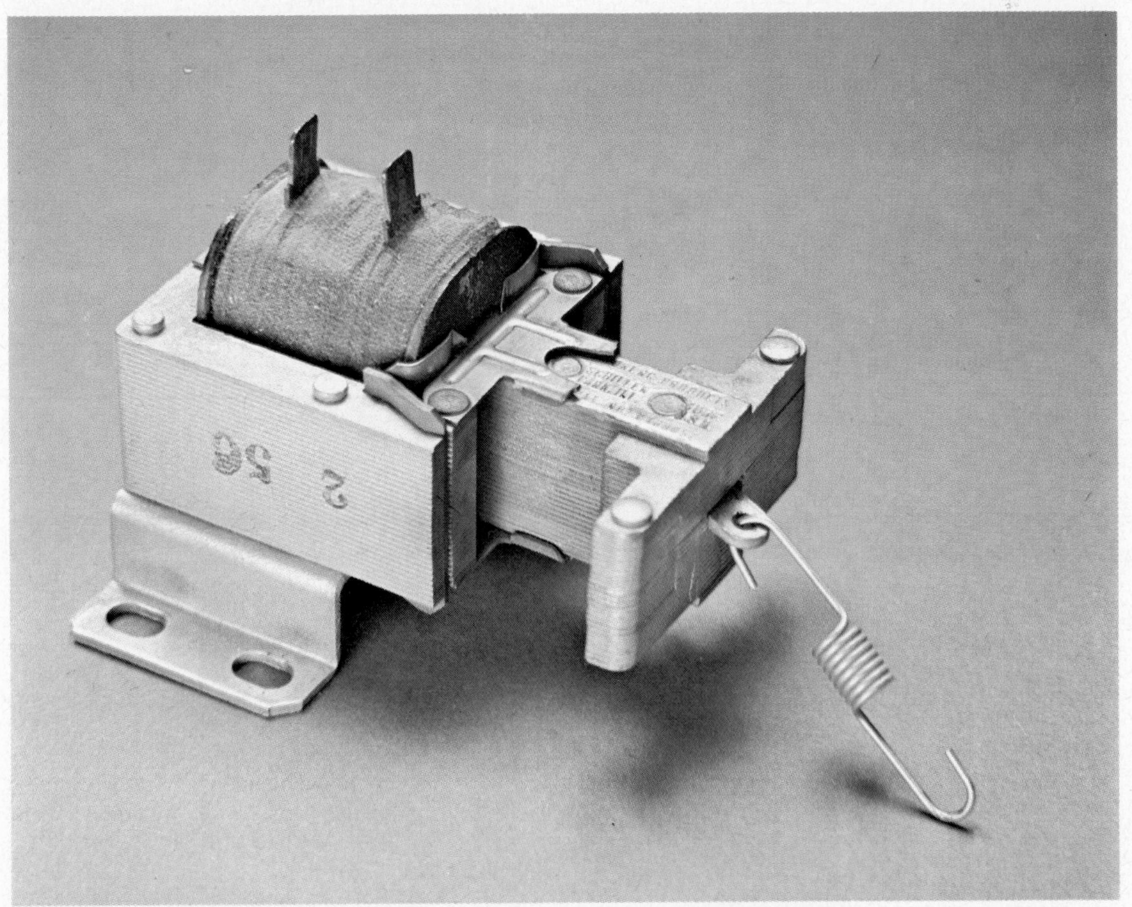

Many heating appliances have thermostats. Volume 10 of this series describes thermostats in detail.

sharply. If this happens too often, the wire may break from too frequent bending. Other obvious problems occur when someone accidentally steps on a plug, shattering it, or accidentally cuts a line cord.

When an appliance fails to work, your first check should be to test the wall outlet, as illustrated in Figure 1-16, and make sure there is voltage at the outlet. If there is not, check your fuse box. If there is voltage at the receptacle, but no voltage across the line at the appliance end of the cord, then the fault is in the cord or one of its connectors. If a cord is broken, it can be replaced. However, if the break is near the wall plug, you can simply cut off the broken end of the cord, and attach the plug to the shortened cord remaining.

A more difficult problem is to find the source of trouble when an appliance sometimes works and sometimes doesn't. This is called an *intermittent fault*, and it usually can be traced to a break in the line cord. Whenever the two severed ends of the conductor are in contact because of the way the cord is bent, continuity will be established and the appliance will work. If the fault is in the line cord, it can be found by switching the appliance to the *on* position and flexing the line cord slowly beginning at the plug end and progressing toward the appliance. If the intermittent fault is caused by a break in the cord, the appliance will operate at some positions of the cord as it is bent, but not at all positions. From this you can determine where the break occurs. If near the plug, proceed as before; if not, replace the cord.

When some of the strands in a cord are broken, but at least one remains intact, current will still flow, but the current through the remaining strands may exceed their rated current-carrying capacity with the result that the wire gets very hot at the point of the break. This heat is caused by current flowing through the increased resistance of the smaller wire and is evidence also of a drop in voltage. The result is lower voltage at the appliance and consequently less heat produced there. Although this would mean an increase in time to toast a slice of bread, or brew coffee, for example, it is unlikely that the average householder would notice this. However, the hot spot on the cord should be corrected as soon as it is detected by cutting off the injured section if it is near the plug or by replacing the whole cord.

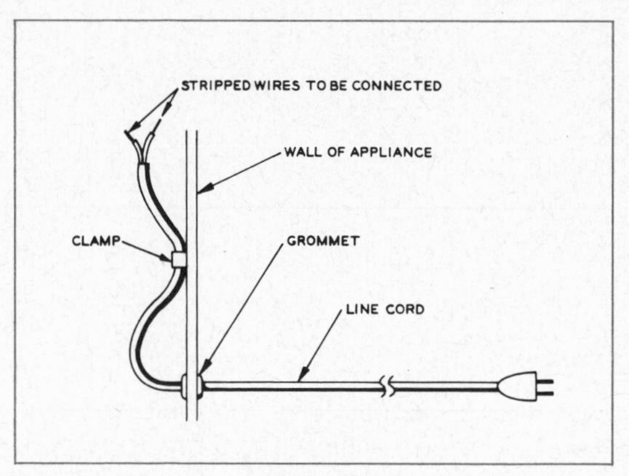

Fig. 3-8. Strain-relief clamp.

When a cord is shortened or replaced, it is necessary to attach a plug to it. It is important to connect a cord to a plug and to an appliance in such a way that the connection will not be subject to strain when the cord is pulled. At the appliance end, the cord is usually clamped a few inches away from the connection, as shown in Figure 3-8. When attaching a line cord to a plug, you can also remove strain from the connections by tying a knot in the line cord after passing it through the connector. The steps involved in replacing a male plug are shown in Figure 3-9, with the typical strain-relieving "Underwriter's knot" shown in (a), (b) and (c). After stripping the wires, slip the plug over the line cord as shown in (a). The first loop of the knot is shown in (b). Slide the knot down to the end of the insulation. (Actually, any kind of knot will do as long as it prevents the cord from being pulled out of the plug.) Pass the stripped wires around the blades of the plug and tighten them under the mounting screws. Be careful not to cross a wire over itself under a screw, or tightening may break it.

If the appliance draws less than two amperes and has a zipcord line cord, you can use the gripper plug shown in Figure 3-6 (c). Here there is no strain-relieving knot, no stripping of wire and no contact screws. Although the connection can be damaged by

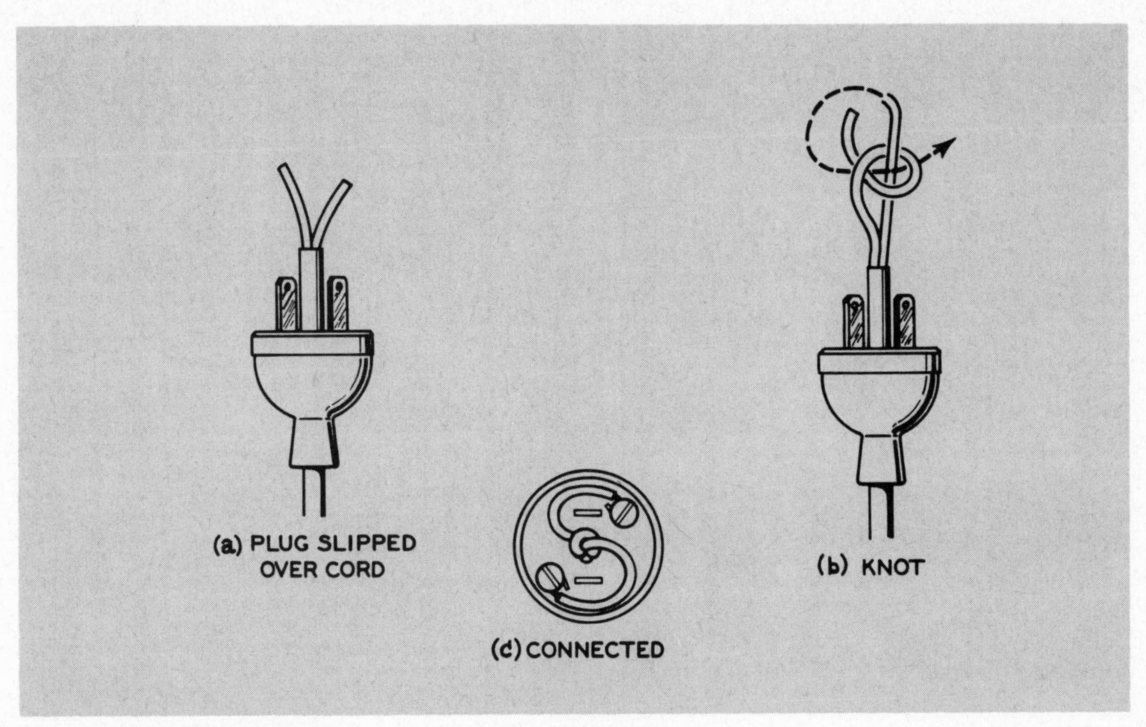

Fig. 3-9. Strain-relieving knot.

jerking the line cord, with reasonable care it can give adequate service indefinitely. There are many varieties of gripper plugs on the market, and all come with simple instructions for installing.

> *WARNING: Do not use gripper plugs on anything but zipcord* and in low-current applications.

Heater cord needs some special treatment. After the heater cord is stripped, as in Figure 3-5(b), there is no longer asbestos between the two conductors in the area outside of the jacket. To remedy this, after stripping and before proceeding further, the asbestos fibers shown hanging over in the figure should be brought up around the rubber insulation of the two leads and fastened there by tying with a piece of thread. Then this cord can be fastened to appliance and plug in the usual manner.

When an appliance plug is to be connected to a cord set, the plug is taken apart by unscrewing the two screws holding it together. If the two halves are riveted together, you might do well to cut off the plug and buy a new one that uses screws. However, it is possible to drill out the rivets and replace them with screws when putting the plug back together again. In either case, inside the plug are two heavy-duty connectors. The two wires of the line cord must be stripped and connected to screws on these connectors. (First, you must slip the strain-relieving spring onto the cord as it will not go over the connectors.) After the wires are connected, the connectors and an end of the spring are placed in appropriate grooves that are cut out to receive them, and the two halves are screwed together. It is *not* necessary to tie a knot in the cord since the interior of the appliance plug is designed to give strain relief without it.

Whenever you have occasion to disconnect a cord at the appliance, note carefully how the cord is routed and where it is connected, so that you can fasten a new cord in the same manner. It is useful to draw a sketch. This is especially important when a part with many connection wires is removed. A sketch of all connections, showing colors of wires, is almost a necessity.

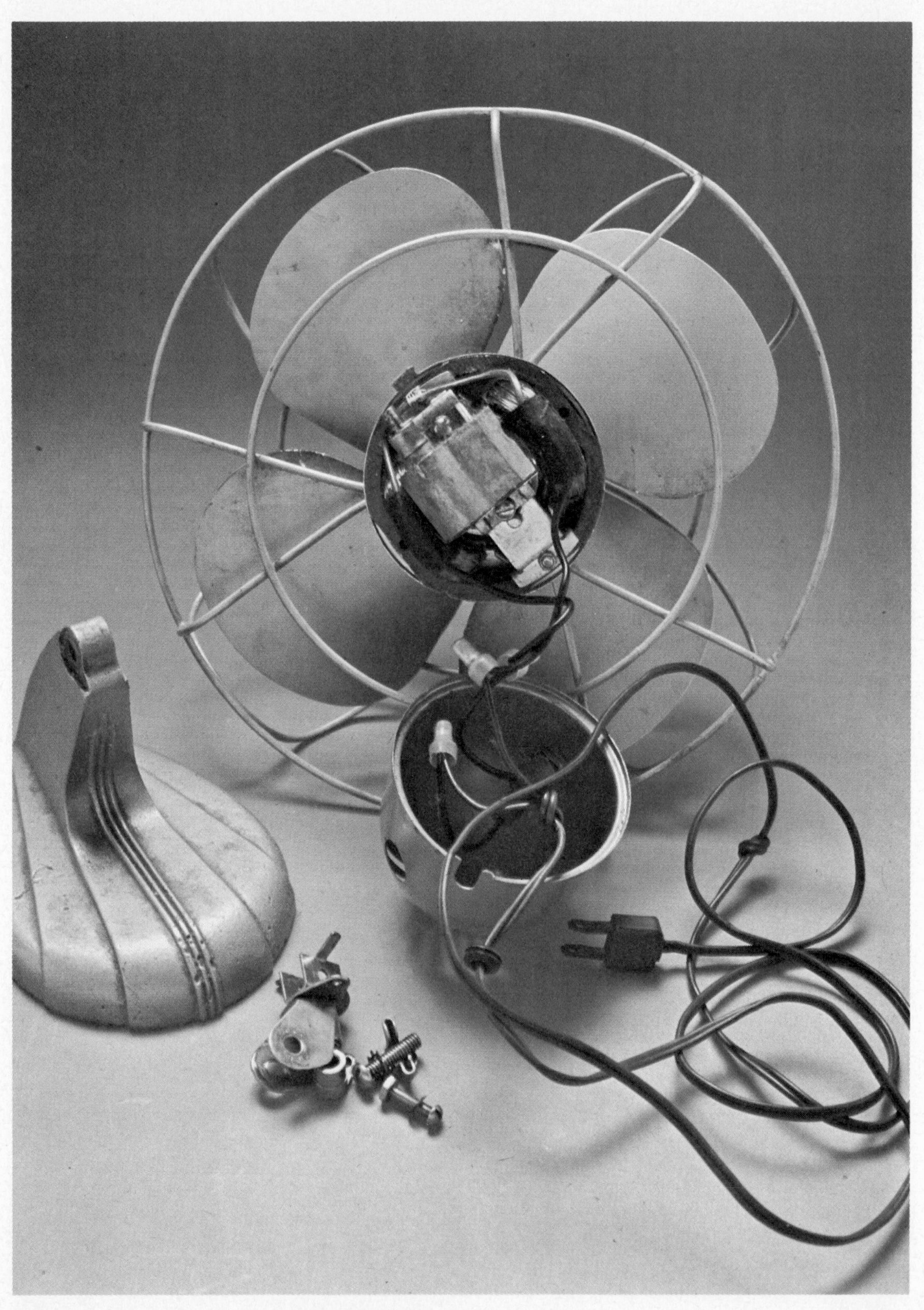

Interior of electric fan showing switch and motor.

Small Appliances with Motors

Many appliances use some sort of motor to move a tool to do a particular job. For example, in a fan, a motor turns the fan blade; in a mixer, the motor rotates the beaters; in a blender, a knife is turned; and in an electric can opener, the cutting wheel rotates. Power tools, dentist drills, and even some toys use a motor for nothing more than rotating or moving a tool, just as in home appliances. In this chapter, some home appliances with motors are considered; their principles of operation and methods of repair can be applied to shop tools and professional instruments as well.

4-1. Electric Fans

An electric fan is a very simple appliance consisting simply of a switch and a motor which turns a fan blade. There are two types: cooling fans for moving air and exhaust fans to remove kitchen odors. They are the same electrically. Fans almost always have an induction motor, as indicated in Table 1-1. Most fans used in the home are less than 1/4 horsepower and use either a shaded-pole or a split-phase motor. Both of these have low starting torque, but air is not a very heavy load. Some fans have a universal motor for use in areas where only direct current is available, but these are rare. Thus, for most fans, brush troubles are nonexistent.

Kitchen exhaust fans are usually controlled by a switch near the stove. This switch connects the outlet to the power line. A sketch of an exhaust fan is shown in Figure **4-1**. The fan is usually mounted in the ceiling or high on the wall above the kitchen range. The fan motor is mounted on two to four metal arms which are attached to the studs by rubber mountings to reduce noise from vibration. A short line cord from the motor is plugged into a single outlet which is controlled by a switch near the stove, as mentioned above. Since this cord is never moved, you never have to worry about line cord troubles in an exhaust fan. At ceiling level, an ornamental grill of metal or plastic hides the fan but allows the hot air and grease to get through. Above the fan there is some sort of covering (not shown in the figure) which remains closed when the fan is not in use, but opens whén it is turning. This keeps out rain and cold. Exhaust fans run at one constant speed.

The motors in most exhaust fans are "permanently" lubricated. This means that they have large wicks soaked in oil in each end bell which should keep the bearings oiled for many years. Nevertheless, after many years of operation, these oil wicks do get dry and the motor needs re-oiling. If the oil wicks are dry, one sympton is that the fan turns too

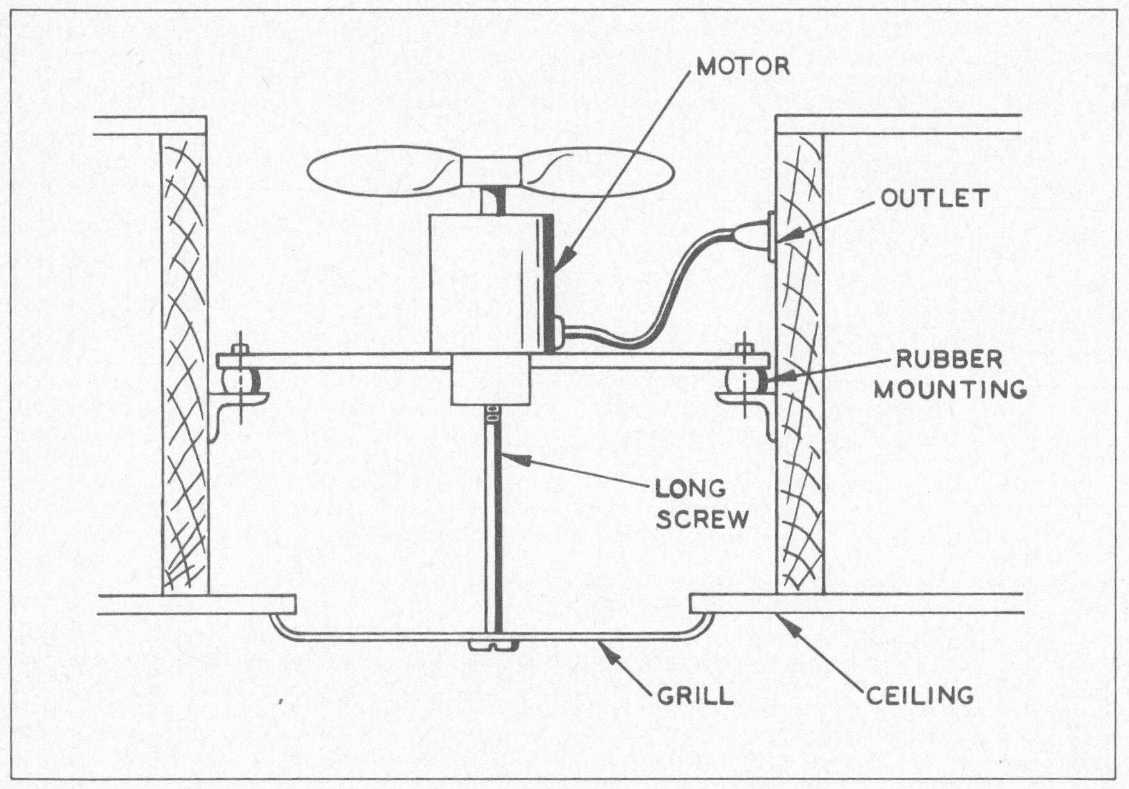

Fig. 4-1. Exhaust fan.

slowly. To check this, remove the grill and spin the blades with your fingers. The blades should coast easily and slowly come to a stop. If the fan stops abruptly or seems to have friction in the bearings, it probably needs oiling. Look for oil holes, and if you are lucky enough to find them, you can oil the fan without detaching it from its mounting. Use light oil and run the motor a few times between successive oilings. If there are no oil holes, you must remove the motor and take off the end bells. Add oil to the cloth wicks until they are saturated, and reassemble the motor. Wipe off any excess oil that has leaked outside the motor.

If an exhaust fan won't run at all, check the motor for continuity. You can do this simply by touching your meter probes to the prongs of the line cord with the switch in the on position. If the circuit is open, the trouble may be in the switch or the field winding of the motor. Usually, however, an exhaust motor fails to run because it is clogged with dirt. Dirt in this case means flour, grease particles, and everything else that is sucked up through the grill when the fan is running. Clean the grill and all parts of the motor with paper towels. If grease has gotten inside the motor, take the motor apart and clean it. A good solution is simply detergent and water. You can use this on the grill, the end bells, the rotor, and bearings, but don't use fluid on the stator windings. If the contraption above the motor is covered with grease so that it fails to open, this will also need cleaning.

After you have cleaned the motor, reassemble it. Make sure you oil the bearings before you put it back on its mounting. It is a good idea to plan to clean a kitchen exhaust fan regularly before waiting for it to develop troubles. Twice a year should be sufficient.

Fans for cooling a room by moving air come in a variety of sizes and shapes. They are usually mounted in a round or square frame to prevent people from sticking their fingers into the whirring blades. Incidentally, if you are holding the blade of a small fan when it is turned on, the blade will not turn because the motor has low starting torque. Don't try this, however, on large fans which use capacitor-start motors. Cooling fans usually have some speed control arrangement.

Most fan troubles are mechanical rather than electrical. The fan is dropped or knocked over and a blade is bent or rubs against the frame. If a fan vibrates or is noisy, a blade may be out of balance from being bent. Rotate the fan by hand and make sure each blade is the same distance from some fixed reference point on the frame. A simple way to do this is to hold a pencil against a bar of the frame so that it just touches a blade, as shown in Figure 4-2. Then hold the pencil steady and rotate the blades. Each blade should just touch the pencil. If necessary, bend the blades carefully until they do so.

If the fan blades hit against the frame, you can bend the frame away so that they turn freely. You should then check as above to make sure the blades are balanced.

If an oscillating fan fails to oscillate, but runs smoothly otherwise, the problem is caused by stripped gears or a worn clutch in the oscillating linkage. If you don't need the oscillating feature, you may decide to do nothing, since the fan and motor still operate

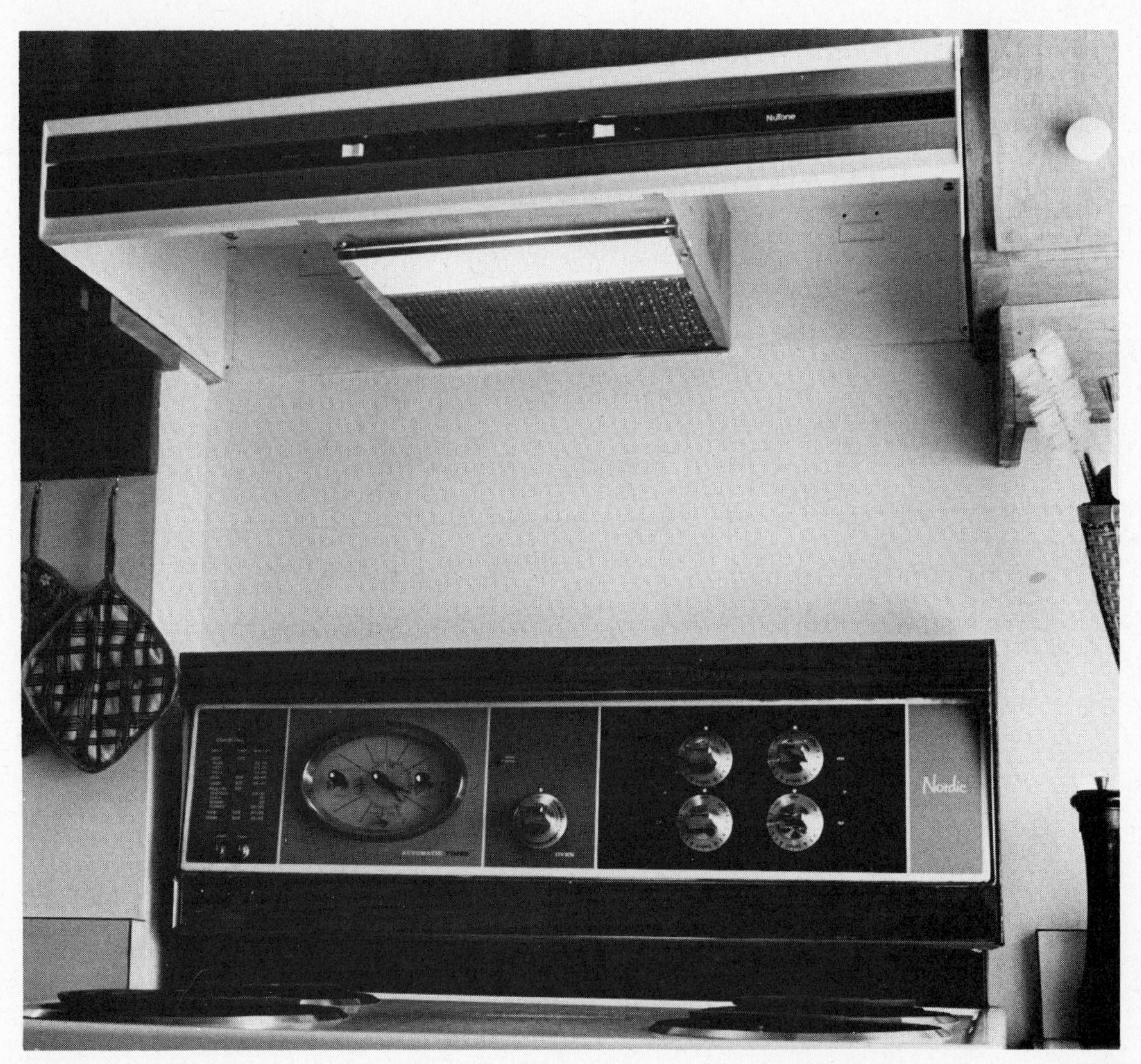

Kitchen exhaust fan.

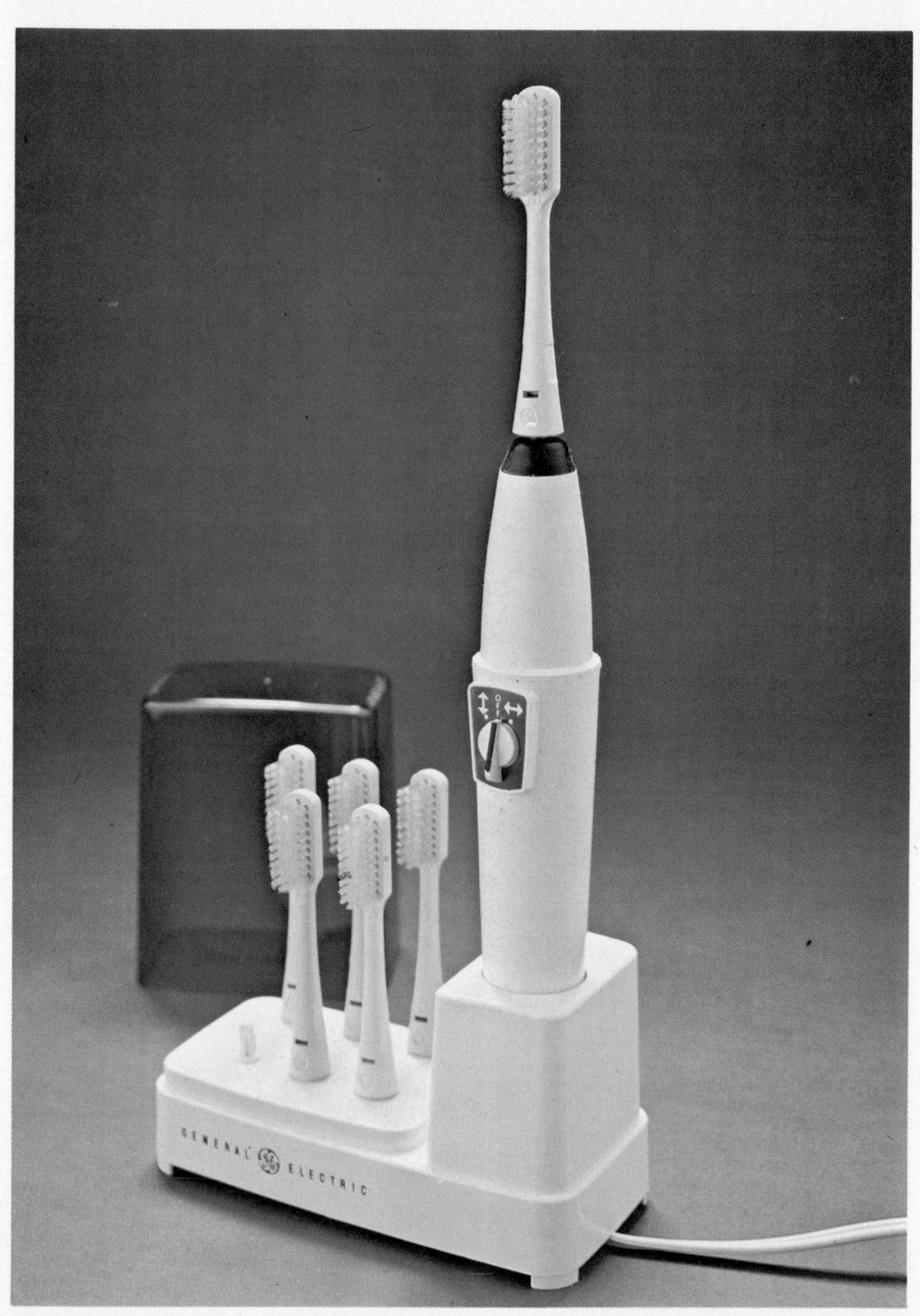

Electric toothbrush.

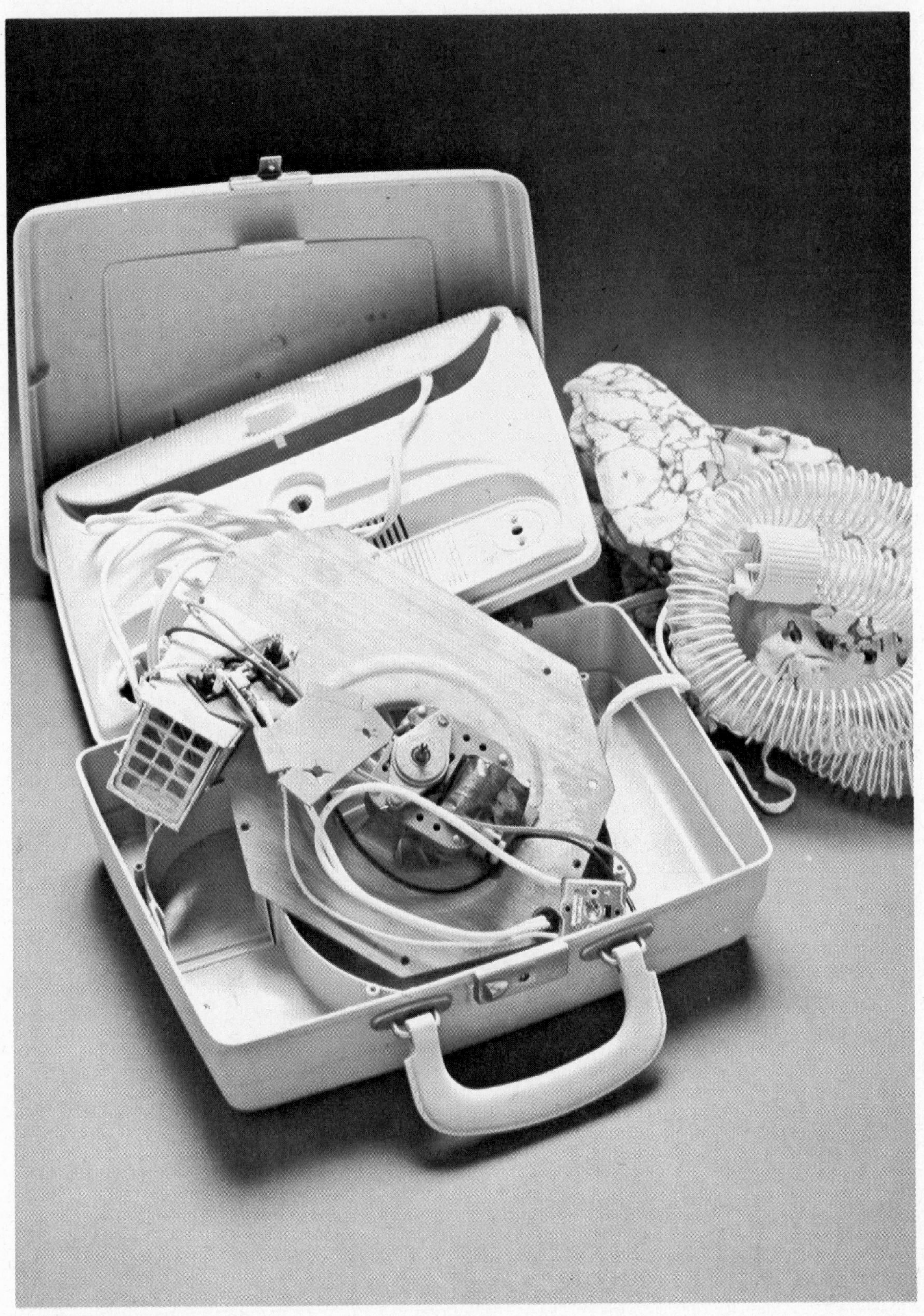

Interior of hair dryer.

Fan-heater.

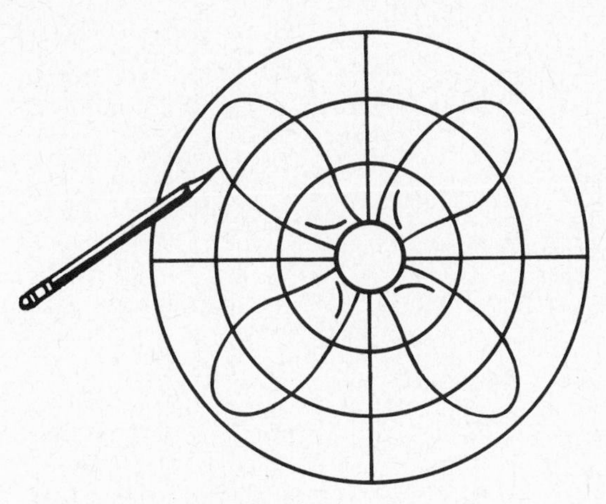

Fig. 4-2. Checking blades.

properly. However, if you wish to repair this, you should replace the whole linkage, since if one part is worn, others will soon wear out. When a new linkage is installed, put a drop of light oil on each pivot point and bearing.

If any type of fan won't run at all, check to see if there is voltage at the outlet first. Then your next check should be a continuity check across the prongs of the line plug, with the switch turned to the on position. Note that this checks line cord, switch, and motor windings at one time. If an open circuit is indicated, then the line cord, switch, and motor should be checked separately. For line cord tests and repairs, consult Chapter 3, Volume 4, and for motor maintenance, consult Chapter 1 in Volume 10. Dirt clogging the fan blades or motor can also stop a fan. Dirt can be washed off blades and mechanical parts with water and a detergent, but the motor windings must be kept dry. Use a wooden stick or an old toothbrush to remove dirt from the motor.

If it is necessary to take apart the motor, you must first remove the fan. The housing holding the blades is usually held on the motor shaft by Allen screws. Loosen these screws with an Allen wrench and slide the blades off. Now the motor is accessible for repair or cleaning.

If the switch in a fan is faulty, remove it and replace it with a similar part. As an interim measure, you can by-pass the switch until you get a replacement. The fan can be used in the meantime, but it will start when it is plugged in; to stop it, you must pull out the plug.

Some large attic fans have a belt drive between the motor and the fan blades. If the fan seems to lack power, it may be due to a slipping belt. There is usually an adjustable pulley for tightening the belt. If the belt itself is too worn, replace it with a new one.

A multiple-speed fan usually has separate field windings, one for each speed. A selector switch connects the electricity to the proper winding for the chosen speed. If the fan runs on all but one speed, the trouble may be a bad contact in the switch or a burnt-out winding. In either case, a continuity test will locate the trouble. Another way to determine the cause of the fault is to move the wires leading to the switch contact. For example, suppose the fan runs at low and high speeds, but won't run at medium. At the switch, swap the low and medium wires so that the low-speed field coil is connected to the medium position on the switch and vice versa. Now if the fan runs at its slow speed when the switch is at medium, but fails to run when the switch is at low, then there is nothing the matter with the switch contacts. The trouble must be in the motor winding. On the other hand, if the motor runs at medium speed with the switch in low position, the medium contacts in the switch are at fault. If the switch is bad, it can be replaced easily. If a motor winding is at fault, the problem is more complicated. In some motors all the windings will have to be replaced since they are wound together. You must first take the motor apart before you can determine what new parts are necessary.

4-2. Juicers

Electric juicers are very simple appliances. A reamer for squeezing the juice of oranges, lemons and similar fruit is mounted on the shaft of a motor. The motor is mounted so that the shaft is vertical and the reamer is at the top of the juicer. The motor itself is a simple split-phase motor which has low starting

Disassembled blender.

Juicer.

torque. This is why the motor will stop if you press down too hard with the orange you are squeezing.

Since the reamer should turn much slower than the motor, a gear box is mounted between the two to reduce motor speed to a suitable reamer speed. In some models, a strainer is added, so that the juice must pass through the strainer before coming out the spout. This strainer may be stationary, it may oscillate, or it may turn in the reverse direction. A reciprocating linkage causes oscillation, and an additional gear is used in models where the strainer rotates backwards. The gear box is usually fastened to the motor with Phillips head screws. At the other end of the motor shaft, a fan is sometimes attached to cool the motor when it is running. An illustration of the electrical circuit of a juicer with case removed is shown in Figure 4-3. Note that the line cord is connected through a switch to the field coil in the motor. If a fan is included, as in the figure, the bottom of the case must have holes in it to let in air for cooling.

Because of its simplicity, a juicer rarely gives trouble. Line cord troubles are rare even after years of service. If a line cord is bad, it can be fixed by the techniques described in Chapter 3 of this volume. The switch is the most likely candidate for trouble. If the juicer fails to operate when the switch is turned on, and if the line cord is plugged into a live receptacle, your first check should be continuity. Remove the plug from the receptacle and measure the continuity between the prongs of the plug, with the switch on and off. When the switch is off, of course, you should get an open circuit.

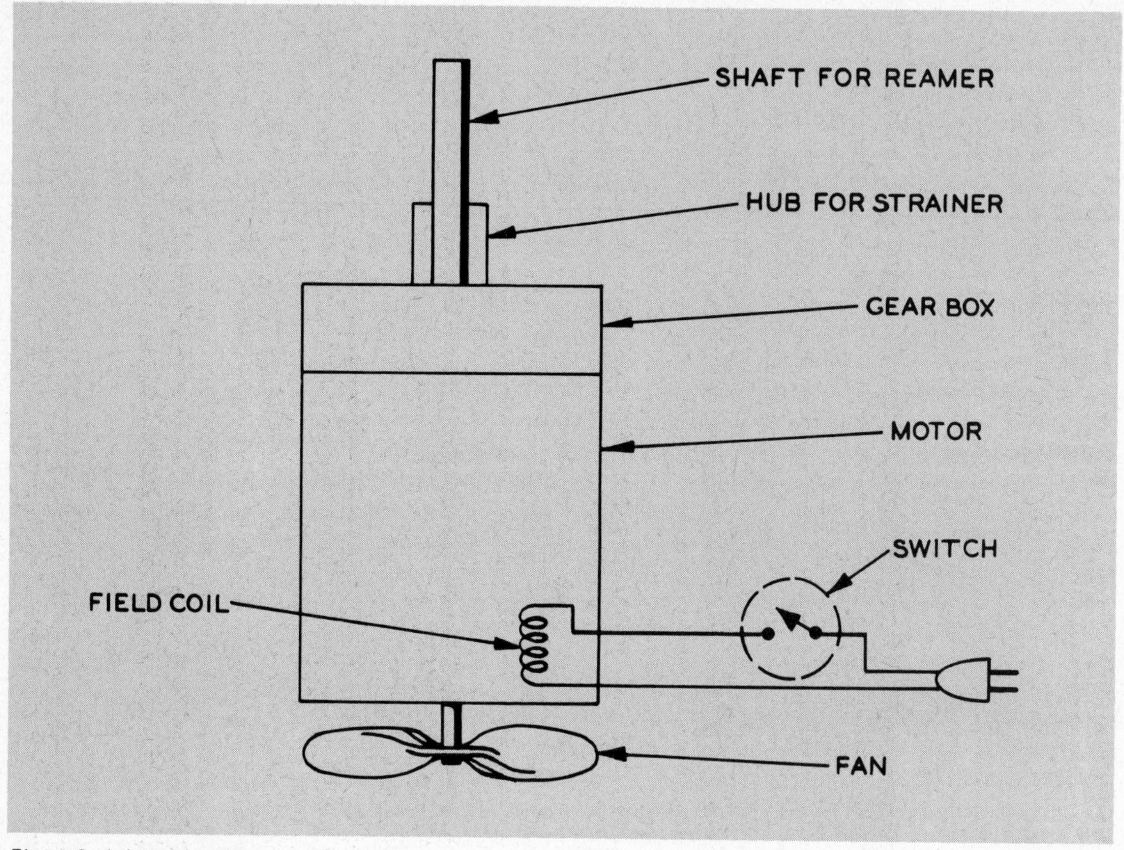

Fig. 4-3. Juicer circuit.

When the switch is connected, there should be a continuous path through the switch and field coils, as can be seen in Figure 4-3. If there is no continuity for either position of the switch, open the case and check directly across the switch and again directly across the field coils. The case is easy to remove, since it is held only by two to four Phillips head screws, and the motor and switch are easily accessible. If the switch is bad, replace it with a similar part.

If a motor winding is bad, you might still read continuity across the terminals, since the starting coil is also in the circuit. For motor troubles and their solutions, refer to Volume 10. The motor in a juicer should give no trouble, however, unless liquid gets into it. Thus, never wash the motor casing, but simply wipe it off with a damp rag or sponge. If, through neglect, the case is badly encrusted with fruit juice or other dirt, it can be removed from the motor and washed in warm water and detergent. If the case is plastic, as many are, it should not be washed in an automatic washing machine.

4-3. Mixers

In an electric mixer, a motor drives two or three beaters for mixing, stirring or whipping a variety of ingredients in the preparation of food. When mixing heavy batters, the beaters must turn slowly, and the motor is heavily loaded. On the other hand, when whipping cream or beating eggs, the load is light, and the motor rotates rapidly. Because the motor must operate under widely varying load conditions, including starting under heavy load, universal motors are used in mixers.

To meet the different speed requirements, some method of varying the speed is necessary. Three different methods of varying speed are described in Volume 10. All three are used in mixers. In general, if the mixer has three distinct speeds which can be selected by moving a switch, the mixer uses a tapped field coil for speed control. If the speed is continuously variable, then the mixer uses either a movable brush or a centrifugal governor.

Thus, the motor in a mixer is versatile. It starts and runs under all sorts of load conditions, and its speed is adjustable. Why not then use it for many other uses as well as mixing? All that would be required would be to remove the beaters and substitute some other tool which can be motor-driven. Since this is indeed easily done, manufacturers do offer an assortment of attachments with their mixers, and in most mixers these are easily placed in position to be driven by the motor. These

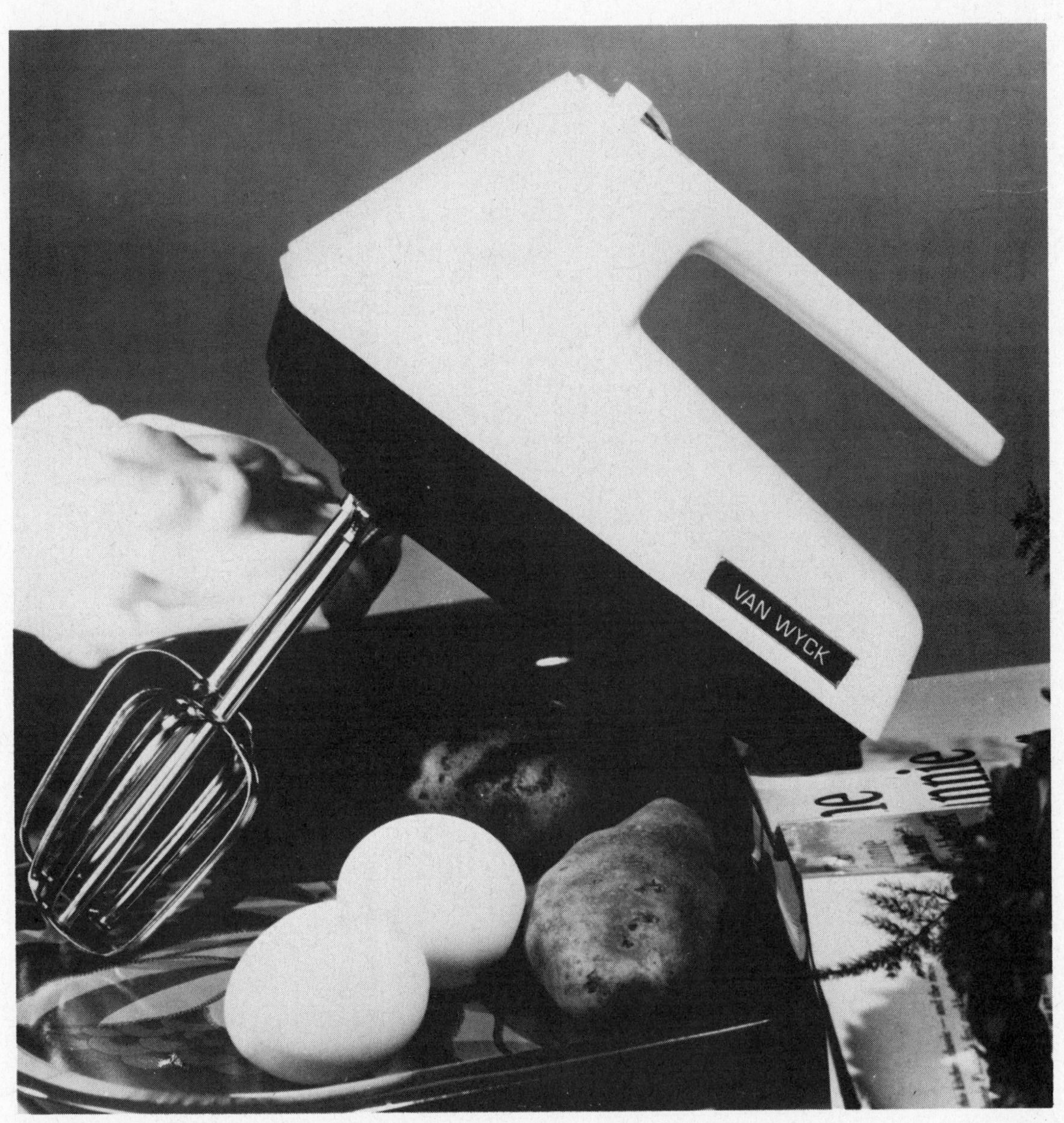

Mixer.

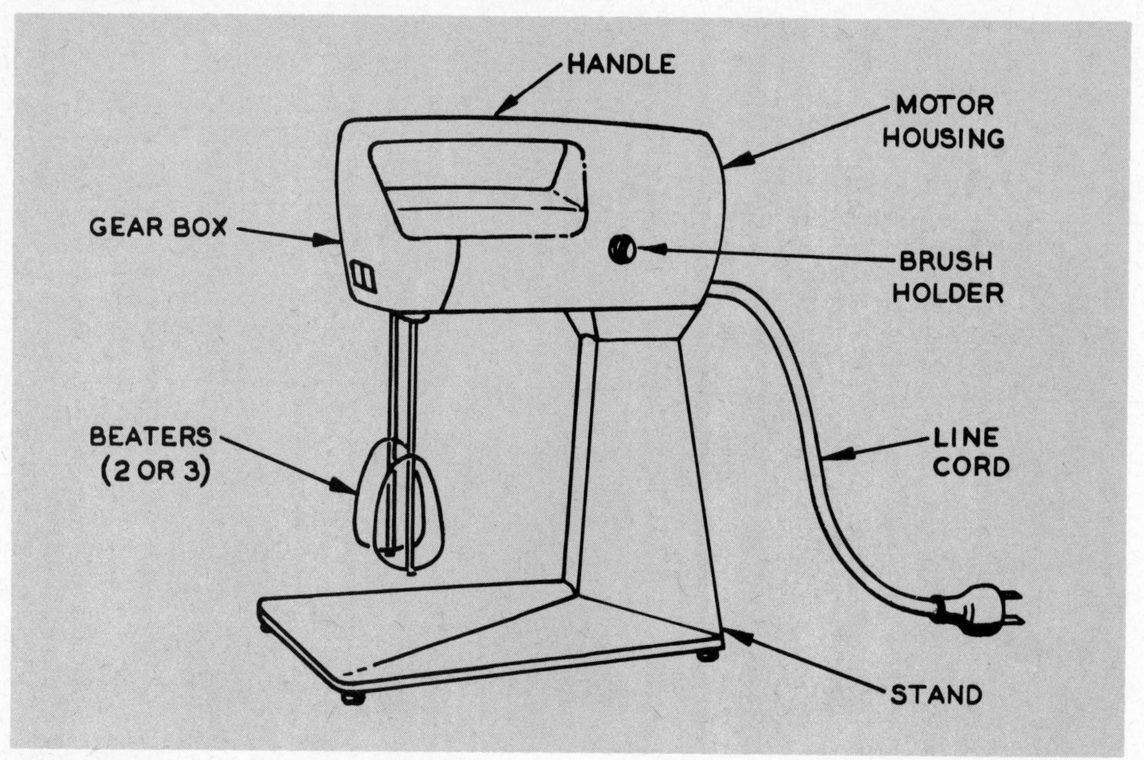

Fig. 4-4. Mixer.

attachments include meat grinders, juicers, potato slicers, ice crushers, knife sharpeners and many others. Despite the attractiveness of having a large group of functions in one appliance, many housewives prefer to use the mixer for its original purpose and consider it too much trouble to change attachments. This section will be limited to a discussion of the mixer without attachments.

An outline of a typical mixer is shown in Figure 4-4. In most modern mixers, the top part including the motor and beaters can be lifted off the stand so that the housewife can carry the mixing part of the appliance to the stove to stir food while it is cooking. Notice in Figure 4-4 that the brush holders are removable from the outside of the case because in a universal motor brushes wear and need to be changed. Thus, it is seldom necessary to open the case, since the brushes can be removed without opening it, and other motor troubles rarely occur.

The motor is mounted with its axis horizontal, and a worm gear mounted on the shaft engages other gears in the gear box to drive the beaters. Some mixers also have a shaft at the top of the gear box for attaching other appliances. The gear box is lubricated with a special grease which is sealed in and should last indefinitely. It is not necessary to oil the motor or other parts of a mixer with normal use.

Most mixers use capacitors to minimize radio interference. Also, those that use a centrifugal speed control have additional capacitors to prevent arcing at the contacts. In some models these capacitors are across the line voltage and grounded through another capacitor at their junction, as shown in Figure 4-5. The ground in this instance is the case of the mixer. If the case is not actually tied to a good ground, there is a voltage of about half the line voltage (or about 60 volts) between the case and ground. When the plug is inserted in the receptacle, the case will be grounded through the ground line for one position of the plug, but will be floating if the plug is reversed. To avoid electric shocks, check for a voltage between the case and ground with the mixer both on and off. If there

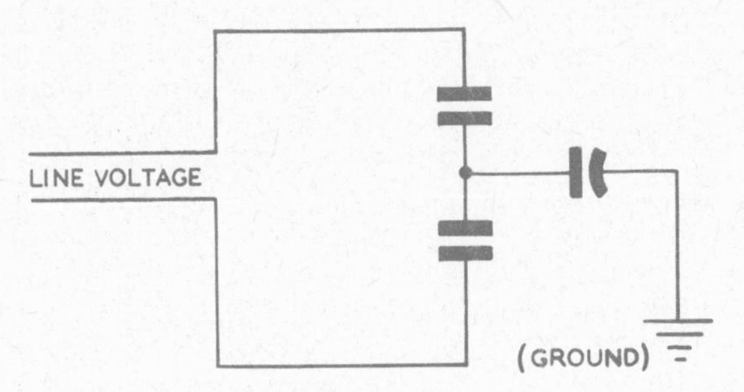

Fig. 4-5. Capacitors across line.

is a voltage there, reverse the plug in the receptacle, and that should remove the voltage on the case.

The most common trouble occurring in mixers is bent beaters due to carelessness or abuse. If the beater blades are bent, you can straighten them, but if the shaft of the beater is bent where it enters the mixer, it may be impossible to make it straight enough to enter without forcing it. You can buy new beaters to replace the old ones.

If the mixer will not run with the plug in a live receptacle, check the line cord and switch, making appropriate continuity tests. Then check the brushes. If they are stuck or worn, the motor will not run. To repair the unit replace the worn brushes. If the brushes stick, you will have to take apart the case and motor to get at the commutator. Sandpaper the brushes and commutator as explained in the chapter entitled "Electric Motors" in Volume 10.

Other things which can prevent the mixer from operating are an open centrifugal switch on the speed control, an open coil in the field or armature of the motor, frozen gears or encrusted foods. Check the centrifugal switch. It should move easily. It may be stuck in the open position due to lack of oil at the pivot point or too much dirt there. Clean and oil it as required. If the contacts are too worn, or the spring is broken, make the appropriate replacement.

An open coil can be detected by a continuity check. This and other motor troubles are described and remedies are presented in the chapter on electric motors.

If the gears are frozen, it will be necessary to take the gear box apart. Clean the individual parts and repack the gear box with an approved lubricant according to the manufacturer's instructions. If this seems like too tedious or difficult a task, simply replace the whole gear box. Note that if the gear box is taken apart, you must observe carefully how

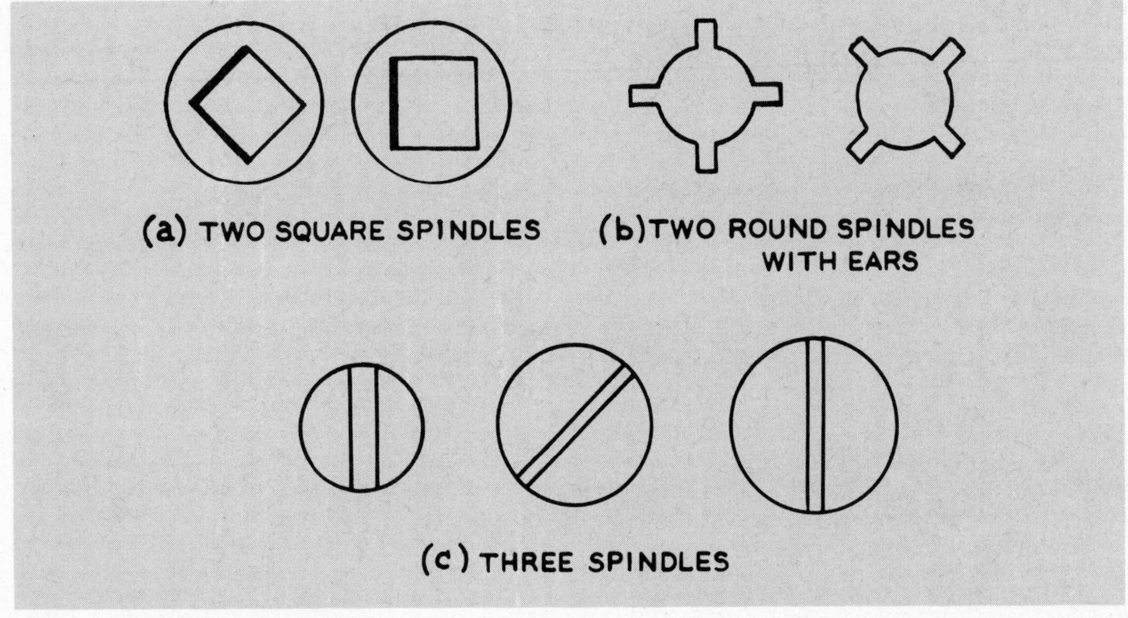

Fig. 4-6. Spindle arrangements.

each part fits. It is worthwhile jotting down the positions of all the parts on a piece of paper as you go along. When the gears are replaced, remember that the beaters must not interfere with one another. The spindle openings or driving slots must be in one of the arrangements shown in Figure 4-6.

If the mixer is encrusted with batter or dirt, it should be taken apart and cleaned. The motor windings must not get wet, but other parts can be soaked in hot water and detergent. Use an old toothbrush to reach hard-to-get-at areas. After cleaning and drying, oil the bearings and reassemble.

If the motor sparks or does not run smoothly, the trouble is either in the brushes or the commutator. Check the brushes and replace them if necessary. If the motor must be disassembled to get at the commutator, it is a good idea to clean it at the same time. Commutator and brush troubles and repairs are described in Chapter 1, Volume 10.

If the mixer runs, but is noisy, it may need lubrication. Check the gear box and lubricate it if required. Other causes of noise are worn bearings or gears and an erratic centrifugal speed control, or governor. Replace defective parts as required. Excessive vibration can also be caused by an erratic governor.

If the motor runs, but the beaters do not turn, the fault may be broken gears or a broken or loose coupling. The pin holding the gear to the shaft may be sheared. Replace all defective parts and make sure new gears are properly lubricated.

If the mixer has no power, it may be dirty or in need of lubrication. Also check contacts in the speed regulator. Clean and lubricate them as required.

If the mixer speed cannot be varied, the capacitor across the contacts in the speed regulator may be shorted. This in effect shorts out the contacts so that the motor is never turned off part of the time, as it should be when the governor is operating properly. Remove the capacitor, and if the mixer now operates correctly, the capacitor is at fault and should be replaced. Note that the capacitor is necessary to prevent too much arcing at the contacts. The mixer speed will be constant if the governor contacts are stuck together. Replace the contacts if necessary; sometimes sandpapering them and oiling the pivot arm will solve the problem.

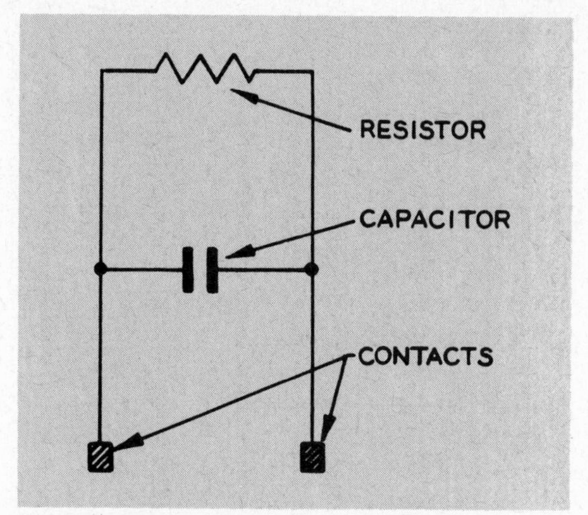

Fig. 4-7. Resistor added across contacts.

In some centrifugal governors, a resistor is added across the contacts along with the capacitor to smooth out the running of the motor. This is shown in Figure 4-7. Now when the contacts are open, as shown in the figure, the resistor presents a path for current, and the motor is not shut off completely. However, because it is a high resistance path, the motor will run more slowly. Thus, the effect of the switch is to change the motor from full torque to a slow torque, instead of from full torque to zero, as it would do without the resistor. This makes for smoother operation. If this resistor opens, the motor may be jerky. In that case, a continuity check across the resistor, with the contacts open, will reveal the trouble. Replace the defective resistor with a new one.

If grease leaks out of the beater sockets, it is a sign of worn bearings or thinned grease. Clean out the gear box and fill with approved lubricant. If trouble persists, a new gear box will be needed.

4-4. Blenders

A blender is similar in principle to a mixer in that a motor drives a tool which acts on the

Blender.

food. In a blender the motor is below the mixing bowl, while in a mixer it is above. Both use universal motors. In general, a blender uses a more powerful motor because chopping and shredding require more work than stirring and mixing. Heavy-duty blenders draw 750 watts of power, but lighter, inexpensive models use as little as 250 watts.

In a blender the cutting knives are permanently mounted in the bottom of the bowl, which may be detached from the base containing the motor, as shown in Figure 4-8. The bowl is held in position on the base by four ears or some similar locating device, and the motor shaft is then coupled directly to the cutting knives. The motor speed is controlled by depressing one of a set of push button switches which connects the line voltage to a tapped field coil. By utilizing only a selected part of the coil for each switch, the desired

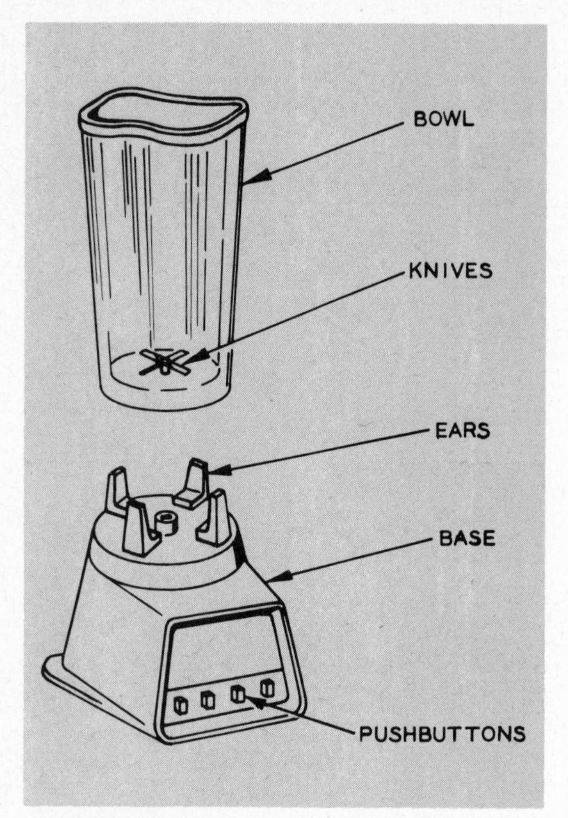

Fig. 4-8. Blender.

Blender-cooker.

speed is obtained. Some older models have a single switch and run at only one speed.

In general, there are two ways to remove the case of a blender to reach the motor for testing and servicing. In some models, a hexagonal nut on the top of the base holds the motor, as shown in Figure 4-9. You can unscrew this nut with large pliers or a wrench, and the motor can then be removed from the bottom. In newer models, you must turn the base over. The bottom of the base is held on by the same screws that hold the rubber feet in place and by two nuts holding the plate to the rotor. These are shown in Figure 4-10. Remove all four feet and the two small nuts, and then you can remove the bottom plate, exposing the motor.

The motor is usually surrounded by a heavy paper cone to protect it from dust and food particles. This should be removed carefully to get at the brushes and other parts of the motor. If the paper cone is torn in removal,

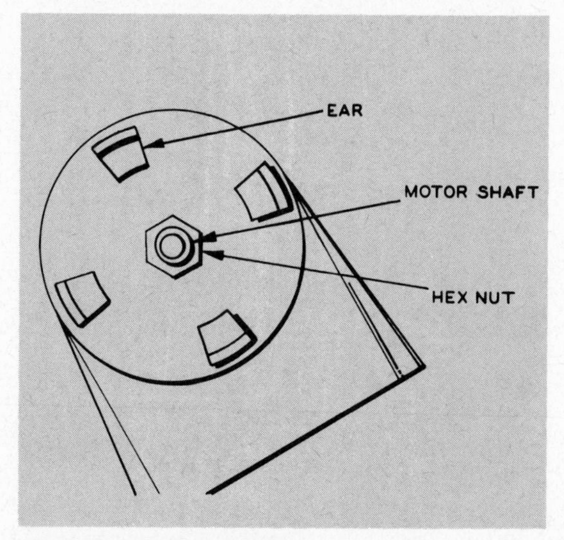

Fig. 4-9. Top view of base.

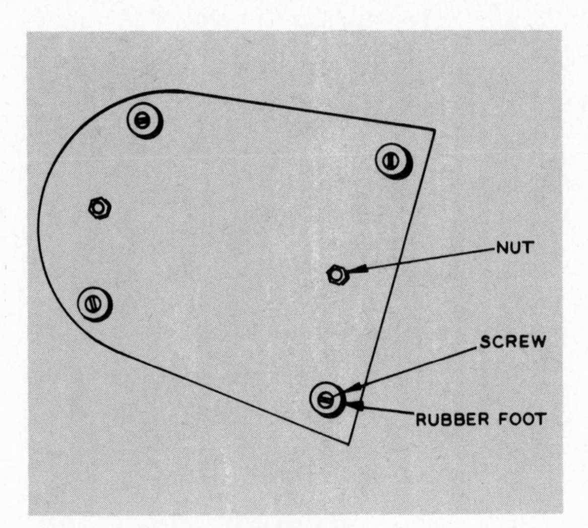

Fig. 4-10. Fasteners at bottom.

you should replace it with a new one when reassembling the base.

If a blender will not run or runs without power when it is switched on, the first thing you must do is determine whether the trouble is in the base or the bowl. Remove the bowl and try again. If the motor runs well without the bowl, the trouble is not in the motor. You may be overloading the motor by putting too much food or items which are too hard to cut in the bowl. Try it with the bowl empty. If the knives still do not turn easily, the trouble is in the bushing. This problem is frequently caused by improper cleaning. Soak the bowl in a mixture of detergent and hot water until you can turn the knives easily with your hand.

The motor itself rarely develops trouble, although brushes may wear. For motor troubles, refer to the chapter on electric motors in Volume 10. For line cord problems, see the chapter on cords in this volume. If the motor runs correctly at some speeds, but fails to run at others, either the selected switch is open or the appropriate part of the field coils is burnt out. Switch trouble is more common. Check each switch by a continuity check, making sure nothing else is in the circuit, and replace any defective switch.

If the motor is taken apart, to check the commutator, for example, make sure that all shims on the armature are replaced in the same positions they occupied originally.

4-5. Electric Knives

An electric knife simplifies carving of meat and poultry. The cutting action is performed by two knife blades arranged in a reciprocating fashion so that they are always moving in opposite directions. The blades remain in contact with each other so that a single cut is made in the food. The blades are removable.

The arrangement of internal parts is indicated in Figure 4-11. The motor is a DC motor, which means it has brushes. A small rectifier converts the AC line voltage to a safe, low DC. The rectifier is a solid-state diode which should never give any trouble. The switch is a push button or trigger on the case which is located conveniently at the finger tips so that it may be held in contact when the knife handle is grasped in the hand. Releasing the switch stops the cutting action. The motor shaft has a worm gear at the end of it, and this engages a pinion gear to which the knife blades are fastened. As the pinion gear rotates, the knives move back and forth in opposite directions. The motor shaft and worm gear are made of hardened steel and never wear out. The pinion gear, however, is usually made of fiber or plastic so that the knife will run more quietly. The pinion gear does wear down eventually and needs replacing.

Possible troubles in an electric knife are:

1. the motor won't run
2. the motor runs, but the blades remain stationary
3. the knife operates too slowly or with reduced power
4. mechanical difficulties such as bent knife blades or broken case or handle.

Some problems can be solved without taking the case apart, and as with all appliance troubles, these possibilities should be checked first. If it is necessary to open the case, you must remove several Phillips head

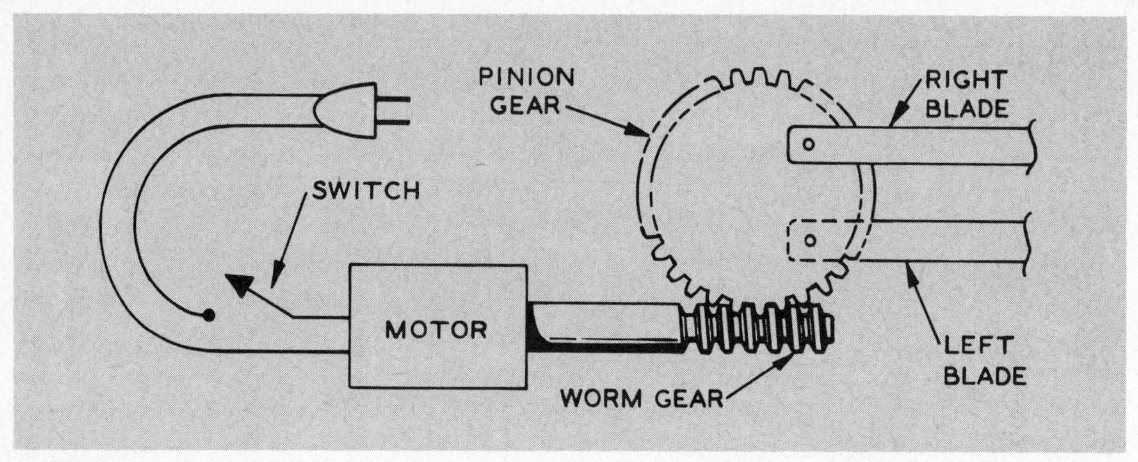

Fig. 4-11. Electric knife.

screws first. The top of the case should lift off easily without force after these screws are removed. It is easy to overlook some of these screws, but all must be removed. With the top of the case off, it is possible to operate the knife and observe what is wrong.

If the motor won't run, check the line cord, outlet, and switch. Refer to the chapter entitled "Line Cords" in this volume for line cord troubles and repairs. A continuity check across the switch will tell you whether the trouble is there. Contacts can be cleaned with sandpaper. If necessary, replace the switch. If voltage is getting to the motor but it still won't run, the trouble may be due to an open coil, a worn commutator, worn brushes or bound gears. Check the brushes first since this is easiest. Refer to the chapter in Volume 10 entitled "Electric Motors" for motor troubles. A continuity test across the motor terminals checks the coils, commutator and brushes at once. Individual parts of the motor

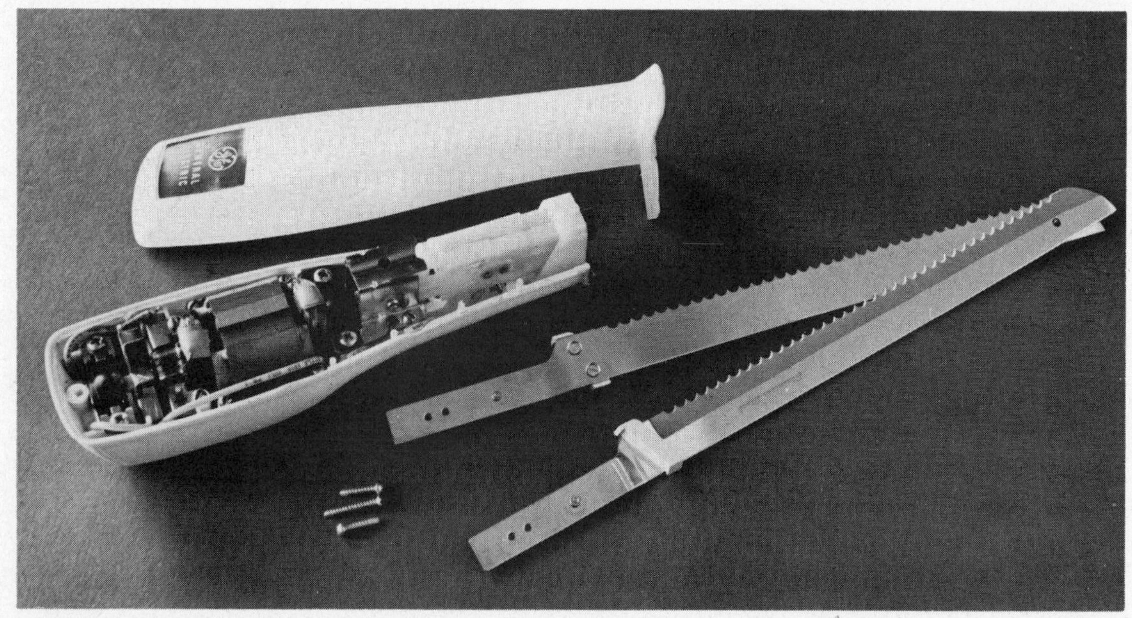

Electric knife blades and case.

can be replaced as required. The most frequent motor troubles are worn brushes or a burnt-out commutator.

If the gears are bound, they may simply need grease or the pinion gear may be warped or broken. Special cream lubricant should be used because it doesn't run and leak out of the handle onto the food being sliced. A broken pinion gear can be replaced. In taking apart the knife, note carefully where each part is located and how it is fastened, and reassemble the parts in the same manner.

If the motor runs, but the blades don't move, the trouble is almost certainly in the pinion gear. Remove it and replace with an identical gear, adding a small amount of cream lubricant as needed.

If the knife is sluggish, the trouble can be due to worn brushes or dry bearings. Check the brushes for wear and also see that the stranded pigtails to the brushes are not broken. If necessary, replace the brushes. The bearings in these motors are made of powdered metal and are impregnated with oil so that they should give trouble-free service. If they eventually do dry up, put *one drop only* of oil on each bearing. This should be sufficient for more than a year's use.

Mechanical problems are obvious. If the case is broken or cracked, it can be fixed with a suitable plastic cement. Bent knife blades should be replaced since they can never be accurately straightened.

Besides the standard electric knife which operates from the house line, manufacturers offer a cordless model which is battery-operated. When it is not in use, the knife is plugged into a battery charger which is always connected to the power line. Thus, the knife is always ready to operate, and on a single charge will run for more than enough time to do all the carving at a meal. The battery-operated model has five nickel-cadmium batteries in the handle to drive the motor, but is otherwise identical to the standard model.

The possible troubles in a cordless knife are much the same as those in the standard model with the addition of battery and charger problems. When the batteries are fully charged, you should measure about seven volts across them without a load; that is, with the knife off. When the knife is running, the battery voltage drops to something between 5.5 and 6 volts. If the battery voltage under load is much below 5.5 volts, one of the cells may be dead or there may be a fault in the charger. Measure the voltage across each cell separately with the motor running. You should read at least one volt across each cell. If one cell is dead, replace all of them.

The output of the charger should be about nine volts. If, in checking the charger, you find no voltage or a low voltage, the trouble may be in the line cord or in the charger itself. Check the line cord to make sure voltage is getting to the charger. If the charger is at fault, replace it.

4-6. Toothbrushes

Like the cordless electric knife, an electric toothbrush also runs on rechargeable batteries. A small DC motor drives the brush in a rapid up-down motion to brush the teeth. The motor is completely sealed in a plastic case to ensure that no moisture enters to damage it. This means that if anything goes wrong with the motor, you cannot repair it. You can, however, send it back to the factory, where the motor will be repaired and sealed in a new case for a small fee.

When not in use, the toothbrush holder is placed in a charger which is always left connected to the house power line. The charger develops about 1-1/4 volts of direct current.

Troubles that can occur are the usual line cord ailments, a bad switch in the charger, or a defect in the motor. Also, if you drop the case, it may crack. You can check for line voltage at the input to the charger with a meter or test lamp. No voltage would indicate a bad switch or line cord, which you can fix. If there is voltage at the input, but no DC voltage at the output of the charger, you need a new charger. Replacement chargers may be obtained separately without buying a whole new electric toothbrush. If the charger is in

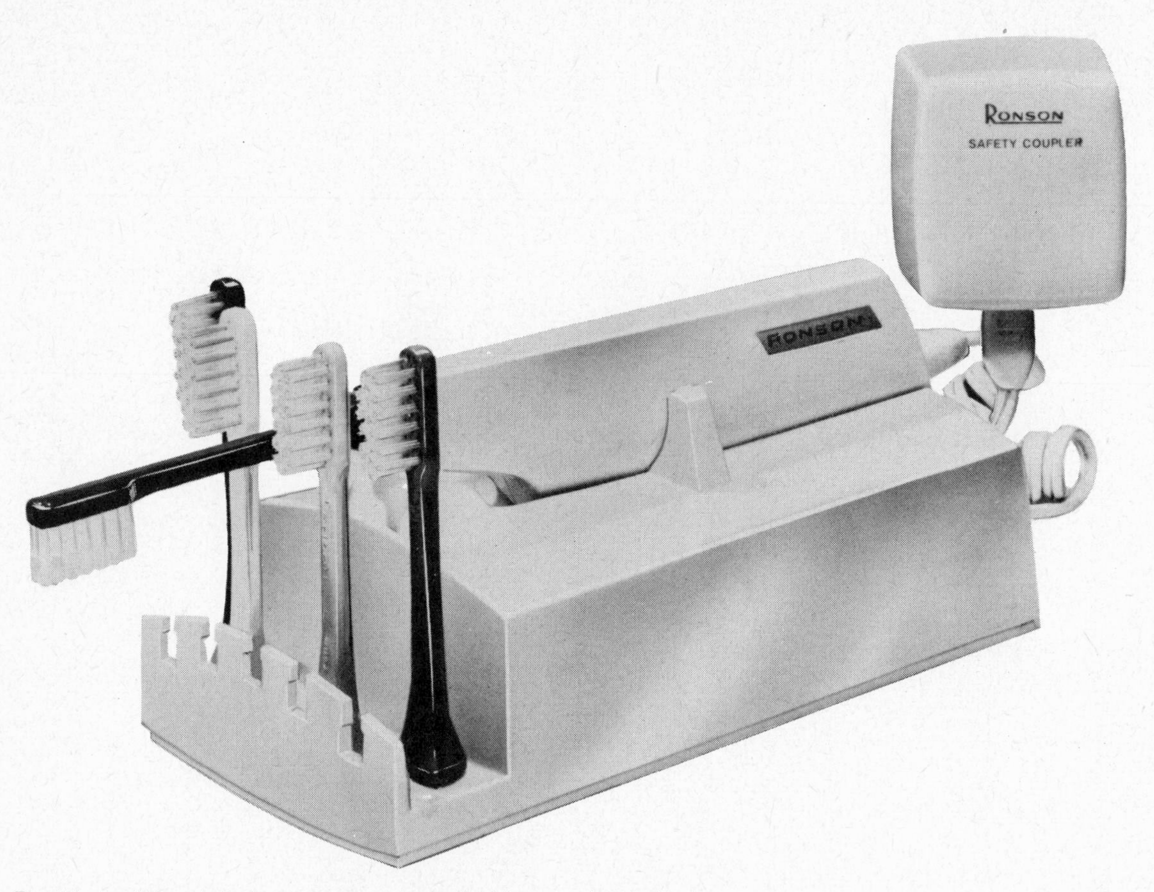

Electric toothbrush that operates on household current or batteries.

working order, but the toothbrush motor refuses to run, you should get a new motor unit, which also can be obtained separately. If the motor runs, but the toothbrush won't move, the trouble indicated is in the gear or coupling. Again, this is something you cannot fix since the unit is sealed. It must be sent back to the factory for repair.

A cracked case can be fixed with plastic cement. However, make sure that the unit is still operative since a fall which breaks the case can also damage internal parts.

4-7. Hair Dryers

The operation of a hair dryer is very similar to that of a space heater. In both, a heating unit heats the air and a motor turns a fan to blow the warm air where it is wanted. A space heater sends the warm air into the room, while a hair dryer directs the warm air to the damp hair which is to be dried.

The parts of a hair dryer are shown in Figure 4-12. The meter is usually a shaded-pole motor which runs on AC only. This type of motor has low starting torque and low power, but since it only works against air, not much power is needed. The shaded-pole motor is relatively inexpensive and rarely gives trouble. The motor turns a fan blade, or impeller, which may be made of metal or plastic. It is usually held on the motor shaft with an Allen head screw.

In most popular models there are two heating elements, typically rated at about 100 and 200 watts. A switch connects either or both across the power line. Thus, for high heat

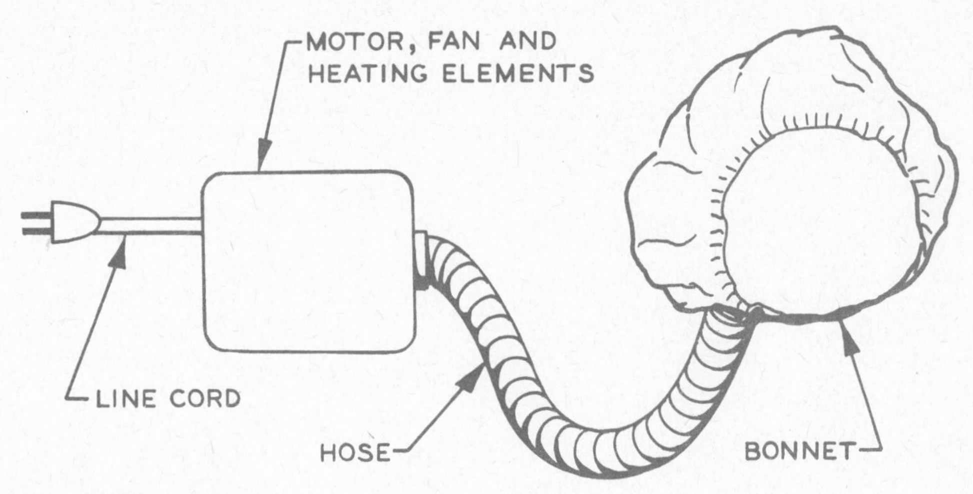

Fig. 4-12. Hair dryer.

both heaters are connected; for medium, the 200-watt unit is used; and for low, the 100-watt alone. The motor is usually connected in the circuit in such a manner that it is impossible to switch on a heating element unless the fan is also running. In most models it is also possible to run the fan alone without a heater turned on so that cool air is supplied.

If the motor had a defect so that it failed to run when it was turned on, then it might be possible to turn on a heating element without having the fan running. This could be potentially dangerous, since the element would be heating the air inside the unit, but the hot air would not be removed. The internal heat could increase to a point where it might cause permanent damage. To prevent this, most hair dryers have a safety thermostat in series with the power line to disconnect the voltage when the internal temperature gets too high. A diagram of the electric circuit of the dryer is shown in Figure 4-13. The thermostat is normally closed. Note that the motor switch and heat selector switch are shown as separate entities, and that the motor must be turned on before voltage can be applied to the heaters. In practice both functions are combined in one rotary switch with five positions. These are:

1. motor and heaters off

2. motor on, heaters off, for cool air

3. motor on, low heater on, for low heat

4. motor on, high heater on, for medium heat

5. motor and both heaters on, for high heat

If an electric hair dryer is completely inoperative (that is, if the fan motor won't run and the heaters do not get hot), there is probably no voltage reaching the unit, since it is unlikely that both the motor and the heaters would fail at the same time. Assuming that the dryer is plugged into a live receptacle, the trouble could be in the line cord switch or safety thermostat. Check the line cord, using the techniques described in the chapter on line cords (Chapter 3), and repair or replace it

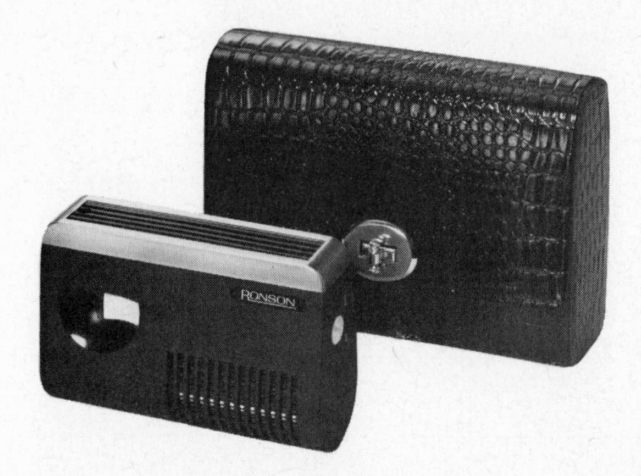

Hair dryer and travelling case.

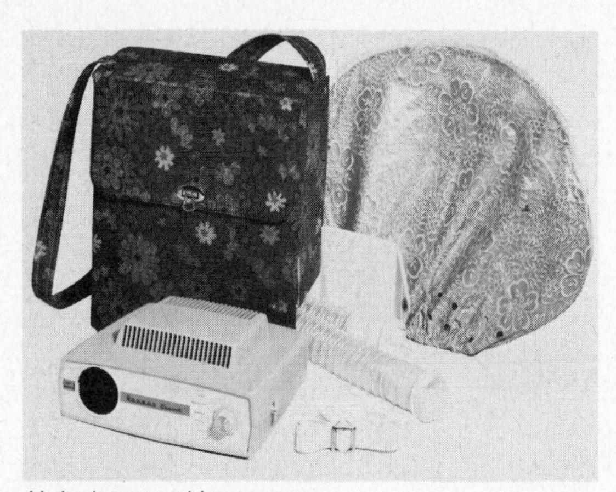

Hair dryer and bonnet.

if necessary. If the cord is in order, next make a continuity check across the thermostat. It should look like a short circuit when cold. *Make sure the dryer is unplugged when you make these tests.* If the thermostat is defective, replace it with one just like it. Note that if the thermostat sticks in the closed position, there would be no indication to the user, since the hair dryer would operate as it should. However, it would no longer have the safety feature. Consequently, if the case of a hair dryer seems to get too hot when the dryer has run for a long time you should check the thermostat to make sure the contacts are not welded together.

If the cord and thermostat are both all right, but the dryer shows no sign of life, the trouble is probably in the switch. You can use a clip lead across the switch to short it out of the circuit, and then the dryer should operate. This would show that the switch is bad and should be replaced. You can check in a more sophisticated manner by measuring voltages at the switch terminals. If the switch is bad and you decide to replace it, make sure you label all wires leading to it so that you can correct them to the corresponding terminals of the replacement switch.

For other troubles, you must first determine whether the fault is in the fan motor or a heater. This is fairly simple. If the heaters light, but the motor does not run, the trouble is in the motor circuit. It may be a fault in the motor or fan, or it may be a bad contact in the switch. Check the switch first. You can test to see whether there is voltage on the motor side of the switch when the switch is on. If not, the switch is at fault. If there is voltage reaching the motor and it still won't run, the trouble may be internal, such as an open coil or a frozen bearing. Refer to the chapter in Volume 10 entitled "Electric Motors". A shaded-pole motor should give no trouble, but if it does have an open coil, it may be simpler to replace the motor rather than trying to fix the coil. If a bearing is stuck, it can be cleaned and oiled. If you take a motor apart to clean the bearings, make sure to note how each part, including shims, is placed, and reassemble in the same manner.

It is possible for a bobby pin or hairpin to come loose and work its way down the hose to the fan assembly, where it can stop the motor. Look for reasons why the fan might be stuck. You should be able to turn the fan freely with your fingers. Noisy operation may be due to the fan hitting its housing or loose impediments. This is fairly obvious and simple to correct. If the fan is bent, straighten it. If it hits the housing, loosen the Allen head screw holding the fan, and move the fan so that it no longer makes contact. Don't forget to tighten the screw.

If the motor runs sluggishly, check the

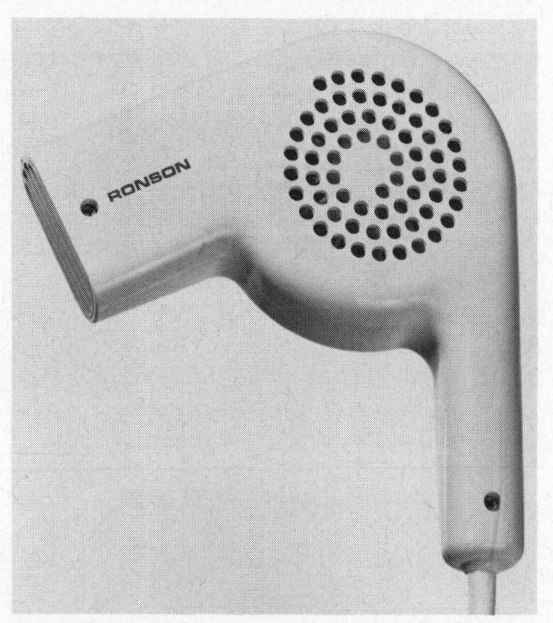

This hair dryer is suitable for home use or commercial use.

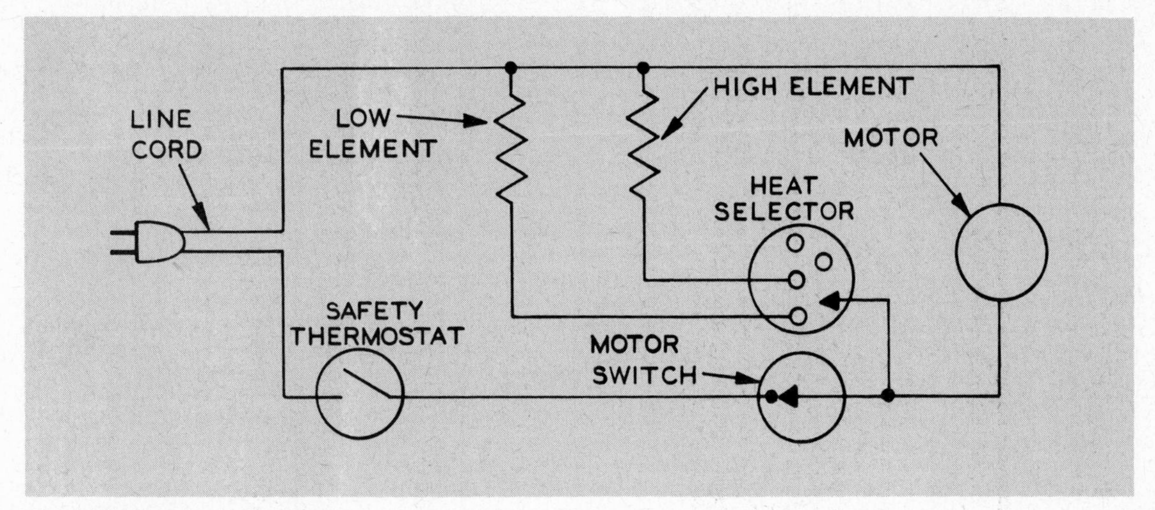

Fig. 4-13. Circuit of hair dryer.

bearings. Clean and oil them as needed. Make sure there is at least 0.005 inch end play in the rotor.

If the motor runs, but neither heater lights, the trouble is probably in the switch, since both heaters should not fail simultaneously. If only one heater fails to operate, the trouble is probably an open heater, but it can also be a poor switch contact. In either case, continuity tests will enable you to locate the defective part. Switch contacts can be cleaned with sandpaper, but it is usually better to replace the switch. A bad heater should be replaced with a new one. Observe carefully how the wires are connected, making a sketch if possible, so that you will fasten the wires to the new part on corresponding terminals.

The air hose shown in the illustration of the hair dryer is made of a wire spiral covered with a tight-fitting plastic sheath. If someone steps on the hose or if it is otherwise abused, it can collapse or the plastic covering may be torn. The best repair is a new hose. However, since this always happens at night or on a weekend when stores are closed, you may have to make a temporary repair. Carefully push a broomstick through the hose and with your fingers bend the spiral wire to support the plastic. It may not look good, but it will be serviceable. If the plastic is cracked or torn, cover the damaged area with plastic electrician's tape.

4-8. Can Openers

Electric can openers work exactly the same as the manual wall-mounted can openers except that the can is turned by an electric motor. An electric can opener uses a small shaded-pole motor to rotate the can under the cutter. Shaded-pole motors have very low torque, and you might suppose they would be incapable of exerting enough force to do the job. The secret is in the gearing. The motor shaft is coupled by a set of gears to a sprocket that turns the can. The gear ratios are such that the sprocket turning the can may rotate at a speed of 1/100 that of the motor shaft. A principle of physics states that the torque is then 100 times that of the motor itself, disregarding friction for the moment. Thus, because high speed is not needed, high torque is available through gearing from a low-torque motor.

In many models of electric can openers, a grindstone is attached to the motor shaft on the end opposite the gears that turn the can opener sprocket. This grindstone then rotates at the speed of the motor and can be used as a knife sharpener. A special slot in the housing of the appliance permits insertion of a knife at just the right angle for sharpening on the grindstone.

An exploded view of an electric can opener with a knife sharpener is shown in Figure 4-14. This is a Toastmaster Model 2210. Note that the motor rotor, part 28, is supported by brackets 25 and 27. The front end of the shaft of part 28 (toward the left in the drawing) has a small gear on it which engages a large gear, part 29. The center hub of part

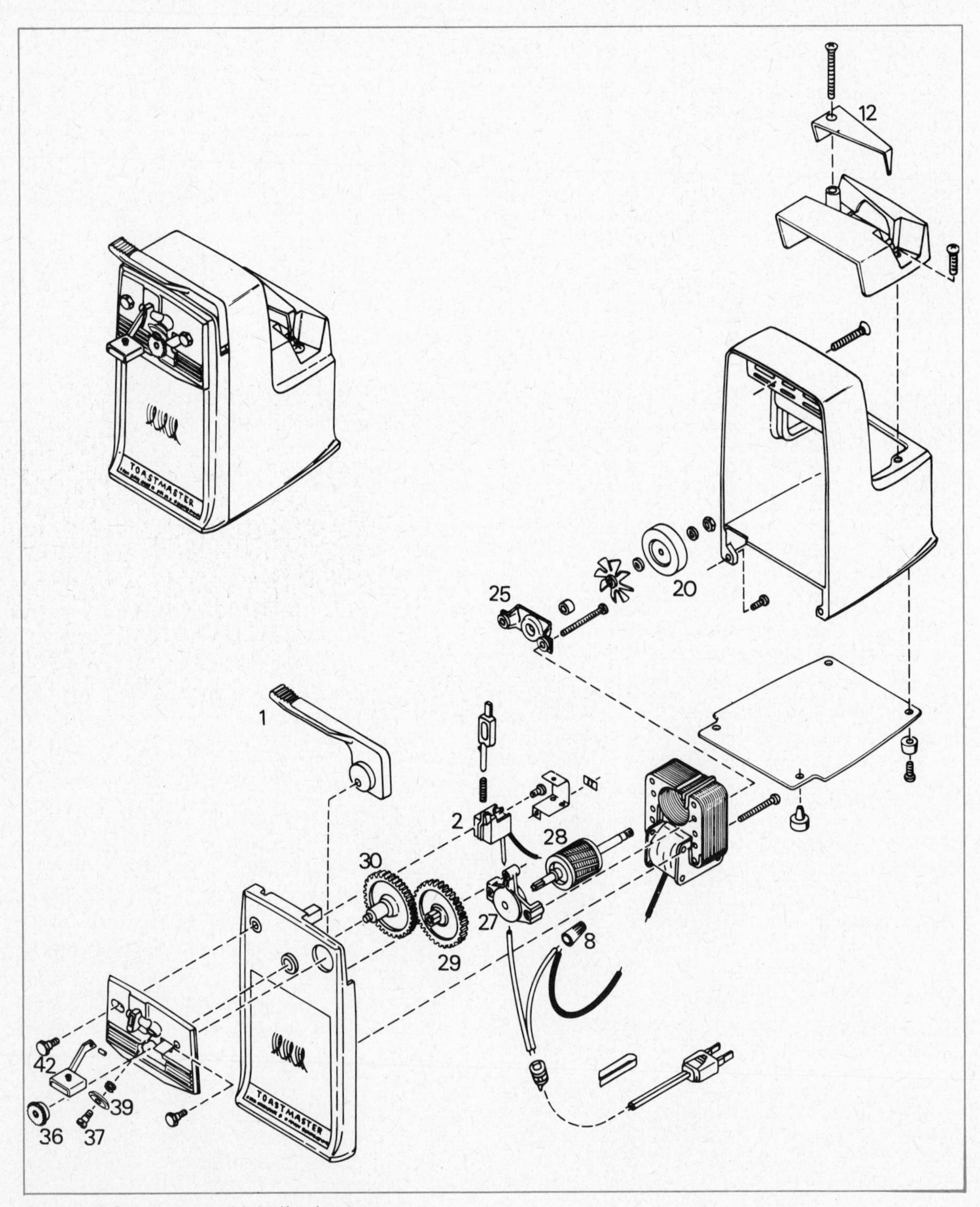

Fig. 4-14. Can opener with knife sharpener.

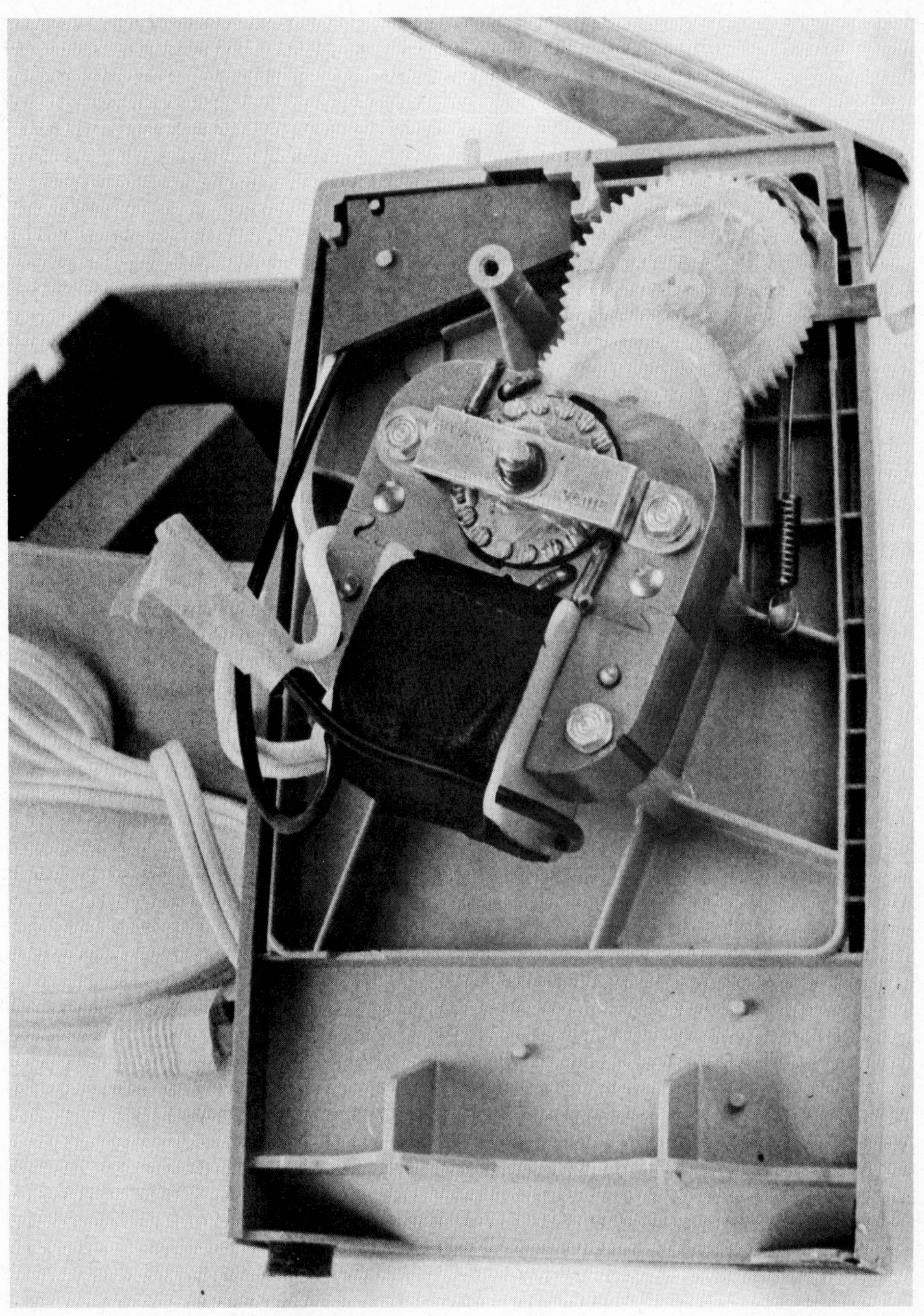

Interior of can opener.

Electric can opener.

29 engages the outside teeth of part 30. The center of gear 30 is connected to the drive sprocket 36. Through the double reduction gears, sprocket 36 turns much more slowly than the rotor and consequently exerts a correspondingly increased torque. The rear end of rotor 28 carries the grindstone, 20, which is used to sharpen knives. Part 12 is a guide to hold the knife at the correct angle against the grindstone.

To use the can opener, the can is placed with its rim against the drive sprocket 36. When the handle, part 1, is depressed, it pushes the cutting wheel, part 39, into the top of the can and at the same time closes the switch, part 2. The magnet, part 42, is pressed against the top of the can to prevent the lid from falling inside when it is completely cut.

If the motor does not run when the switch is on, first check the line cord. Note that one side of the line cord is connected to a motor lead by a solderless connecter, part 8. This cap can be removed to check the voltage at the motor end of the line cord. Check the continuity of the switch also. If the switch is bad, replace it with a similar unit. The shaded-pole motor should give no trouble, although it may need occasional oiling at the bearings.

If the can opener is noisy, the gears or bearings or both need lubrication. To take the case apart, look for Phillips head screws around the case and remove them. The case should come apart without force, when all these screws are removed. Grease can be used on the gears, and light oil on the bearings. The gears will also be noisy if they are too worn. Worn gears should be replaced. It may be necessary to remove the motor to get at the gears, but you should not have to disconnect any wires. Notice carefully how

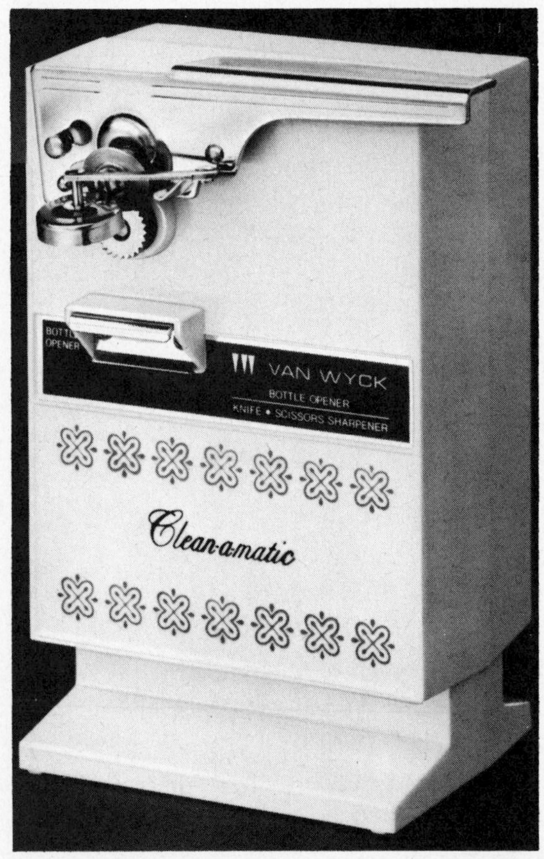

This opener also sharpens knives and scissors.

the parts are arranged, so that replacements may be reassembled in the same manner.

A common fault is that the can opener fails to cut all the way around the top of a can. This can happen if the carrier screw, part 37, is loose. However, a more likely reason is a worn drive sprocket or a dirty cutting wheel. Replace a worn sprocket if needed. If the cutting wheel fails to spin easily, it may be encrusted with food from months of service. It should be washed at least every six months and lubricated with light oil. If the cutter appears dull, you should replace it. Defective parts are easy to remove, but note carefully how they are positioned so that new parts are assembled in exactly the same way.

If the small gear teeth on the end of the rotor shaft are worn, you can replace the shaft. It is not necessary to replace the whole rotor. Before you remove the old shaft, however, measure how far the shaft protrudes from each end of the rotor. When you install a new shaft, make sure it protrudes the same amount from each end. Be careful not to lose any of the shims and spacers which are used to locate the rotor in the correct position in the stator. Look for them when removing the rotor and note their exact positions. When reassembling the motor, return these shims to their correct positions.

4-9. Vacuum Cleaners

Vacuum cleaners may be small, hand-held, portable units or large industrial models. They come in a variety of shapes, usually variations of the three basic types shown in Figure 4-15. The upright model, shown in (a), is self-contained, while the tank and canister types have separate hoses and a set of attachments for doing a variety of housecleaning tasks.

The upright cleaner has a rotating brush in the base which agitates the rug being cleaned so that the dust is released and more easily sucked up into the bag. The brush also helps pick up hair, thread, pins and other small objects on the rug. The nozzle at the bottom of the upright vacuum cleaner is fairly large since it must be wide enough to hold the brush. In effect, this tends to reduce the suction power of the motor, since it is distributed over a wider area. Nevertheless, because of the brush, the upright is the best model for cleaning rugs. The brush is driven by a belt coupling it to the motor.

The tank and canister types do not have a brush, and manufacturers of these types try to compensate for the lack by using a more powerful motor and a small nozzle. Although inferior for rug cleaning, these models do a good job on hard floors such as wood, linoleum or tile. They can also perform tasks not possible with the upright, such as cleaning walls, drapes or furniture. Sometimes both the tank and canister vacuum cleaners are referred to as tank types, with the model shown in Figure 4-15 (b) called a "horizontal tank", and that in (c) a "vertical tank". Either type may have skids or wheels.

All vacuum cleaners use the same principle of operation and essentially consist of two systems, one electrical and the other pneumatic. The electrical system consists of a universal motor, a line cord and a switch. The uprights use a smaller motor, sometimes as small as 1/4 horsepower, while the tank-type

A canister vacuum cleaner.

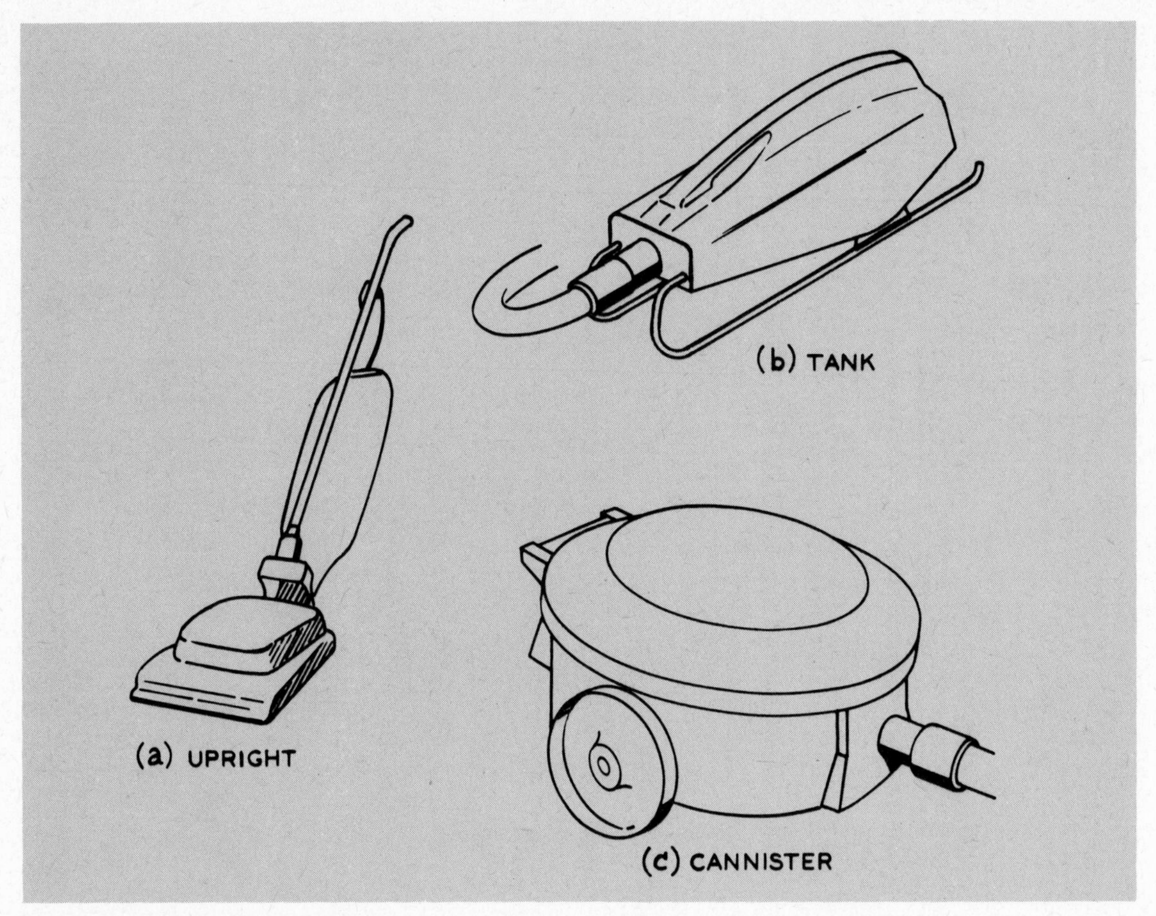

Fig. 4-15. Types of vacuum cleaners.

motors may be 3/4 horsepower for home use, and larger still for industrial vacuum cleaners. In uprights the switch is usually a sliding device on the handle. In tanks and canisters, the switch is usually a foot-operated switch or a toggle switch, but occasionally slide switches are used here also.

The pneumatic or air system sucks air in from the outside, passes the air through a bag to collect the dirt and moves the air outside again. What is required, then, is a *continuous* air passage through the bag. This means the bag must be porous so that air can go through, but of a fine enough mesh to trap the dirt. The air path is slightly different for the uprights and the tanks. In Figure 4-16, the air path for an upright vacuum cleaner is illustrated. A large fan attached to the motor sucks air and dirt into the air intake. The intake, of course, is the nozzle, which is pushed over a dirty surface. The dirty air is blown back through the outlet into the porous bag on the outside of the upright. The dirt is trapped in the bag, but the air passes through the pores in the bag back to the outside. The direction of air motion is indicated by arrows in Figure 4-16. Note that air does not pass through the motor, but flows around it.

The air path in a tank or canister vacuum cleaner is shown in Figure 4-17. Here the bag is positioned at the air intake. Dirty air is sucked into the bag. Again the dirt is retained, but the air continues through the porous bag, past the motor, and back to the outside.

In the upright model, the bag must be emptied periodically and eventually loses its porosity when dirt cakes in the holes. In the tank types, the bags are usually discarded when they are full of dirt and are replaced by new ones which are quite inexpensive. In both

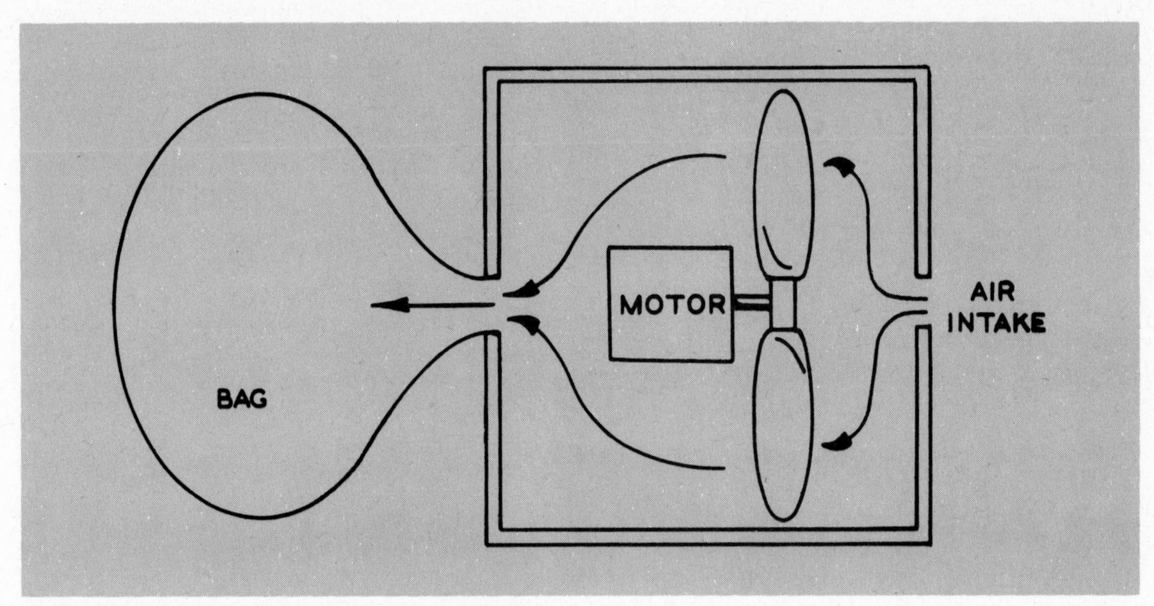

Fig. 4-16. Air path in upright vacuum cleaner.

types, the motor is usually protected from the air flow by a paper or cloth shield to prevent dirt from getting into the motor. Note that in Figure 4-17, if the bag functioned perfectly to filter out the dirt, only clean air could reach the motor. However, some dirt does leak through the pores in the bag and consequently the motor does need protection.

Troubles can arise either in the electrical or air systems, but you can easily determine where the fault is. If the motor fails to run, the problem is in the electrical system. If it runs smoothly, but there is no suction, look to the air passages. If the motor runs sluggishly, however, it may be a fault in the motor, or it may be caused by a block in the air passage, making it more difficult for the fan to pull air through.

Whenever there is a sign of trouble, try running the motor with the air passages open. In an upright, remove the bag and lift the nozzle off the floor. In a tank or canister, open the case and remove the bag so that the motor and fan are free. In either case, the motor should run easily when the switch is turned on. When the switch is moved to the off postion, the motor should coast to a stop gradually. If it stops too abruptly, the cause may be worn brushes or dry bearings. In the case of the upright vacuum cleaner a tight belt to the rotary brush or lint around the brush bearing can also impede the motor. In rare cases, dirt reaching the motor after many years of service may cause the motor to labor. For motor troubles and their cures refer to the chapter on electric motors. If no voltage is reaching the motor, check the line cord and switch. Make sure the cord is plugged into a live outlet. For remedies to line cord problems refer to Chapter 3 of this volume. Continuity checks across the switch will reveal a fault there. Replace a bad switch with a similar unit.

The most common complaint about vacuum cleaners is that although the motor is running, the vacuum cleaner does not clean as it did when it was new. The uprights differ from the others in possible causes for this trouble. For the uprights, first check the rotary brush. As the upright vacuum cleaner is used, this brush is constantly rubbing against a dirty rug and eventually gets worn down. After a while the bristles no longer reach the rug, so that even though the brush is revolving as it should, the vacuum cleaner has lost the "beating" action which the brush is supposed to provide. To take care of this wear, the position of the brush is adjustable, and all it may need is to be moved closer to the bottom of the vacuum cleaner. If the brush is completely worn, you can easily remove it and replace it with a new one. It is also possible that the brush is not turning because the belt driving it is loose or broken. Replace a frayed or broken belt. Tighten a loose belt by moving the brush.

If the brush and motor are in good shape, an upright vacuum cleaner can lose power from a clogged air system. Usually this means too much dirt in the bag, but, as was pointed out above, eventually the pores of the bag get encrusted with dirt, and the bag needs replacing. Try running the vacuum cleaner without the bag, and if it works well, as indicated by a strong flow of air at the post where the bag is usually attached, look to the bag for the trouble. If there is little air pressure there, the fan might be loose on the motor, or there may be a wad of dirt inside the housing. Make the appropriate correction.

Note that about the only possible location of a leak in the air system of an upright is at the point where the bag is attached to the housing. But a leak here will not prevent the vacuum cleaner from functioning. It will simply mean that some of the dirt is blown back into the room instead of passing into the bag. You can check for the possiblility of leak here by listening very closely or by observing the cleaner when it is running.

Tank and canister types can have the same sorts of motor troubles or a loose fan blade, but no trouble from a rotary brush since there is none. If the complaint is poor suction, it may be due to a leak in the air system or a plug of dirt. In Figure 4-17 only the body of the tank housing is shown. A long, flexible hose is connected to the intake, and one of a variety of nozzles is attached to the hose. Leaks can occur at the junctions of the different parts and also in the flexible hose itself. Check the tank without the hose. When you place your hand over the intake hole you should feel strong suction. If not, perhaps the bag is too full of dirt. The most common cause of poor suction is dirt clogging the hose. To correct this, attach the hose to the outlet port of the tank so that air will blow through it in the reverse direction. Usually the air will blow out the dirt and correct the difficulty. If the hose is broken so that leaks reduce the suction, buy a new hose.

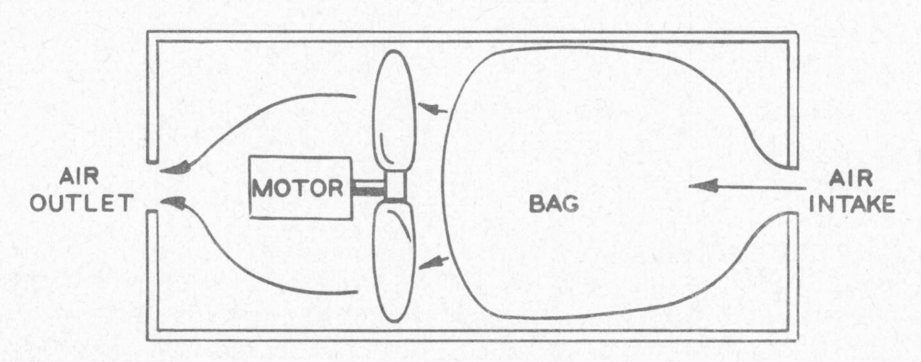

Fig. 4-17. Air path in tank vacuum cleaner.

Small Heating Appliances

The main purpose of the electrical portion of many home appliances is simply to produce heat. The heat may be used for cooking or warming food, as in a toaster or electric skillet, or for warming the person or clothes, as in an electric blanket or electric iron. In any case, the electrical circuit of the appliance is very simple, consisting of a line cord, a heating element, a thermostat and sometimes a switch. The operation of many of these appliances and the repair techniques used when they fail are presented in this chapter.

Before attempting to repair any appliance you should be familiar with the basic principles of appliance repair and with the two basic tests explained in Chapter 1. It may be worthwhile rereading these sections, since in this chapter and those that follow it will be assumed that you are familiar with these principles and tests.

5-1. Heating Pads

Heating pads are used to apply local heat to various parts of the body, usually to relieve pain. The pads may be as small as a foot square and range to two or three times this size. They must be flexible in order to conform to the shape of the joint or other part of the body which needs warmth. The heating pad is usually encased in a washable cover. The pad itself contains two heating elements, one typically rated at 20 watts and the other at 40 watts. These are flexible, thin wire elements which are sewn right into the material forming the pad. The two elements wind back and forth in the pad in parallel. At one end they are joined together and a *common* lead from the junction runs to the control box. The free ends of the elements are each connected directly to a separate terminal on the control box.

A typical control box for a heating pad is shown in Figure 5-1. Note that a three-wire line runs from the control box to the pad, and a two-wire line is used as the line cord. These lines are usually made of zipcord. One lead of the two-wire line cord goes right through the control box and is the common lead for the two heating elements. The other lead runs to the center contact of a rotary switch. The horizontal bar shown at the center of the control box in the figure is used to rotate this switch. With the bar as shown, the switch is open, and thus the heating pad is off. When the switch is rotated counterclockwise so that one end of the bar points to L, an arm of the switch makes contact with the wire leading to the *low* heating element. The circuit for this element is closed, and current flows through it, producing heat. Rotating the bar so that it points to M moves the contact from the *low* element to the *medium* element, producing approximately twice as much heat. In the *high*

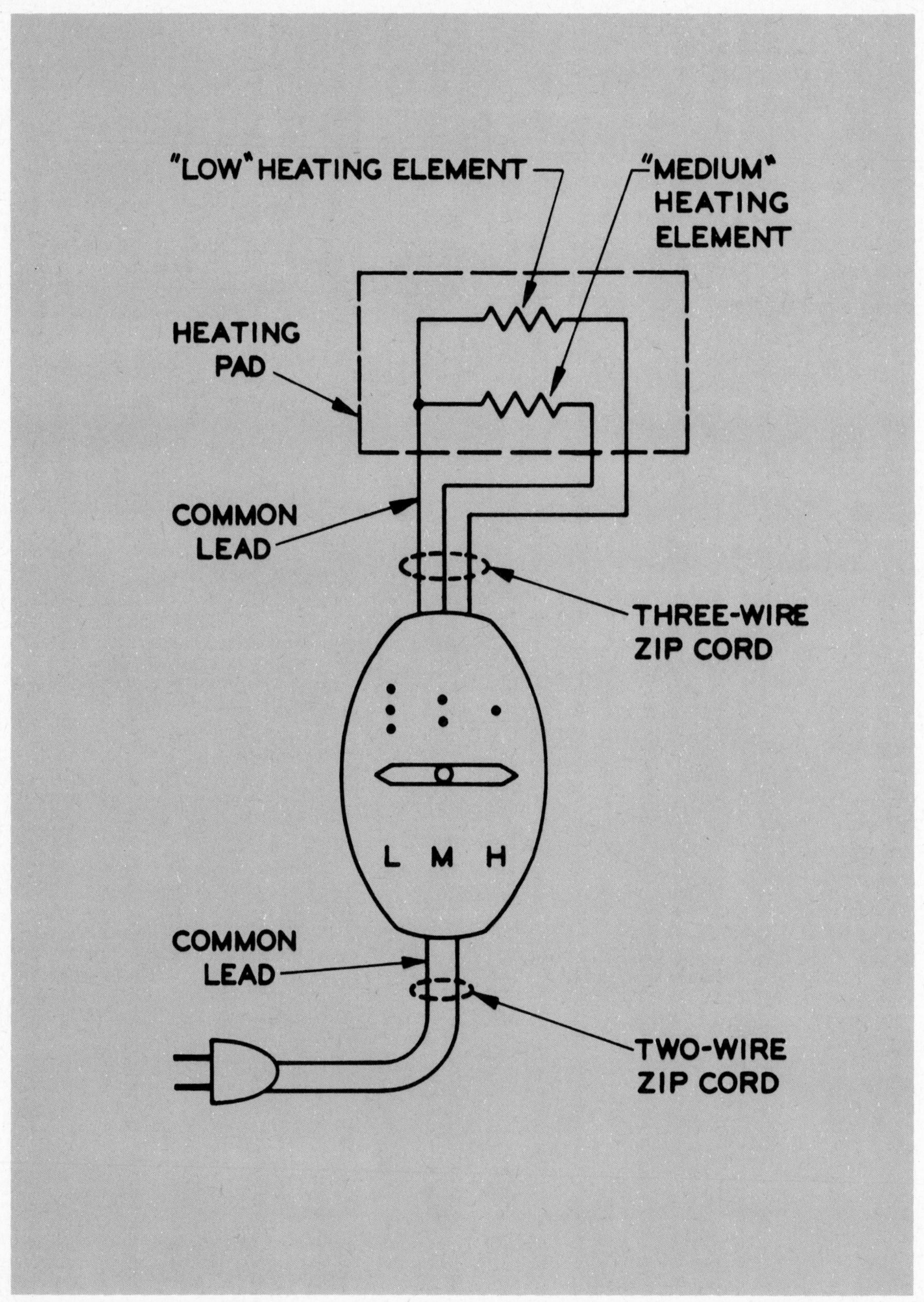

Fig. 5-1. Heating pad control box.

position, when the bar points to H, both elements are connected simultaneously. The dots shown on the control box are embossed on the case so that they may be felt in the dark. This makes it possible to set the control "by feel" without looking at it. When one end of the bar points to L, the other points to a single dot. Two dots correspond to M, and three to H.

The control box is held together by one or two screws which can be removed to gain access to the ends of the two-wire and three-wire zipcords. Voltage checks at the junctions in the box can be used to determine if the line cord is all right and if the switch is operating properly. Alternatively, the switch can be tested by continuity checks. A continuity check is also used to determine whether there is a break in a heating element.

If the heating pad will not get warm in any position of the switch, the trouble is probably *not* in the heating elements, since it would be surprising for both to fail at the same time. Check the switch and the line cord. Also check the connections of the wires to the switch. If the wire in the line cord running to the center contact becomes disconnected, for example, the circuits will remain open with the switch in any position. If all connections are satisfactory, make a voltage check in the control box to make sure the line cord is satisfactory. No voltage means a faulty line cord or a dead outlet.

If the fault is in the line cord or is due to a break in the three-wire zipcord, it can be repaired by splicing in a small piece of wire joining the two ends of the break. This is permissible, since this appliance draws low current. If possible, solder the extra wire in place at both ends, but if no soldering iron is available, a good twisted splice will do. In any case, cover the joint well with insulating tape to prevent danger of shock from bare wires.

If the fault is due to a break in a heating element inside the pad, there is little you can do to fix it. The heating pad may still be serviceable using only one heating element, so it should not be discarded, but you should not try to repair it.

In some heating pads, a small thermostat is placed inside the pad itself, in series with the heating elements. If this thermostat fails in the closed position, it should be ignored, and the pad can be used in the same manner as those without thermostats. If the thermostat fails in the open position, you may be able to remove it or replace it. It is necessary to rip open a seam of the heating pad and find the thermostat. Usually, the bad thermostat is removed completely, and the two wires which were connected to it are simply tied together and *carefully insulated* before sewing up the seam again.

5-2. Electric Blankets

An electric blanket is very similar to a heating pad in that it contains flexible heating elements sewn into the material. There are usually two heating elements, one for each half of the blanket. As with the heating pad, a three-wire line runs from the control box to the blanket. This line usually has a connecting plug and receptacle on it so that the control box can be separated from the blanket. A typical circuit is shown in Figure 5-2.

In a single-control electric blanket, both heating elements are connected simultaneously by one control and the temperature is regulated by one thermostat which is inside the control box. In a dual-control blanket,

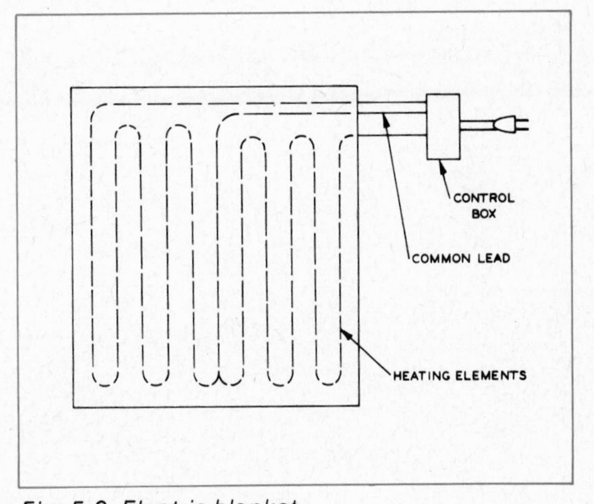

Fig. 5-2. Electric blanket.

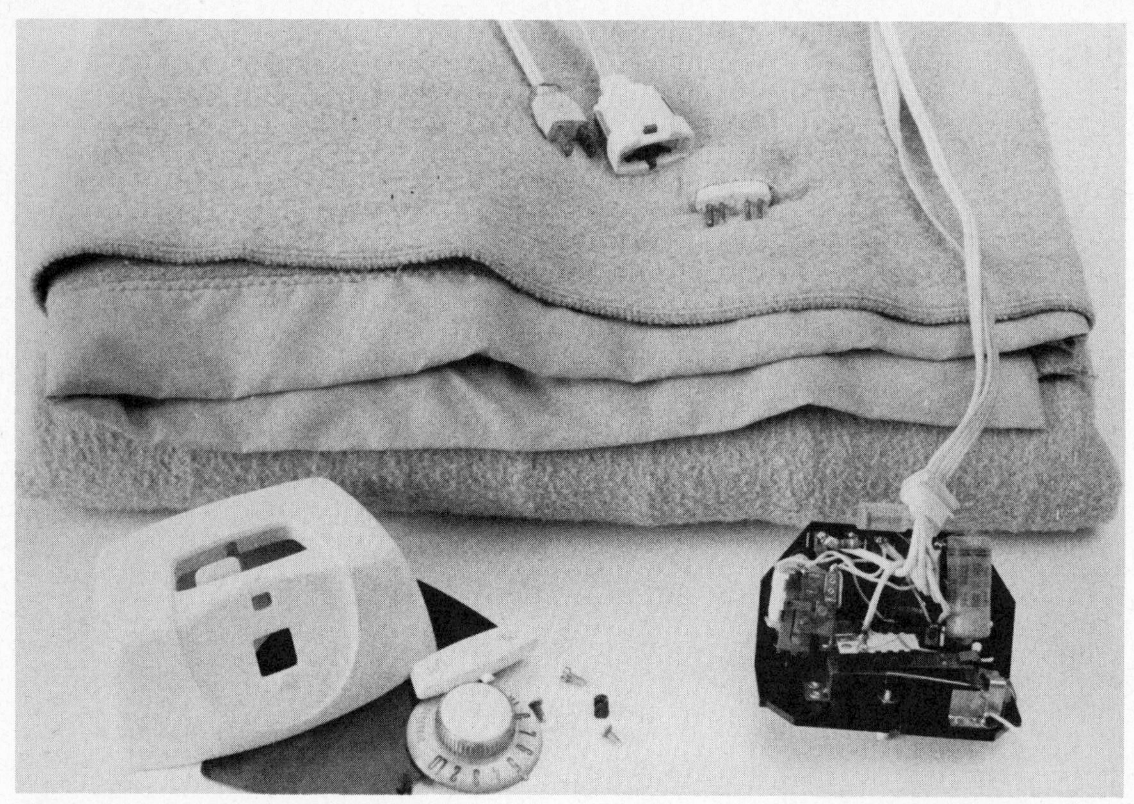

Interior of control box of an electric blanket.

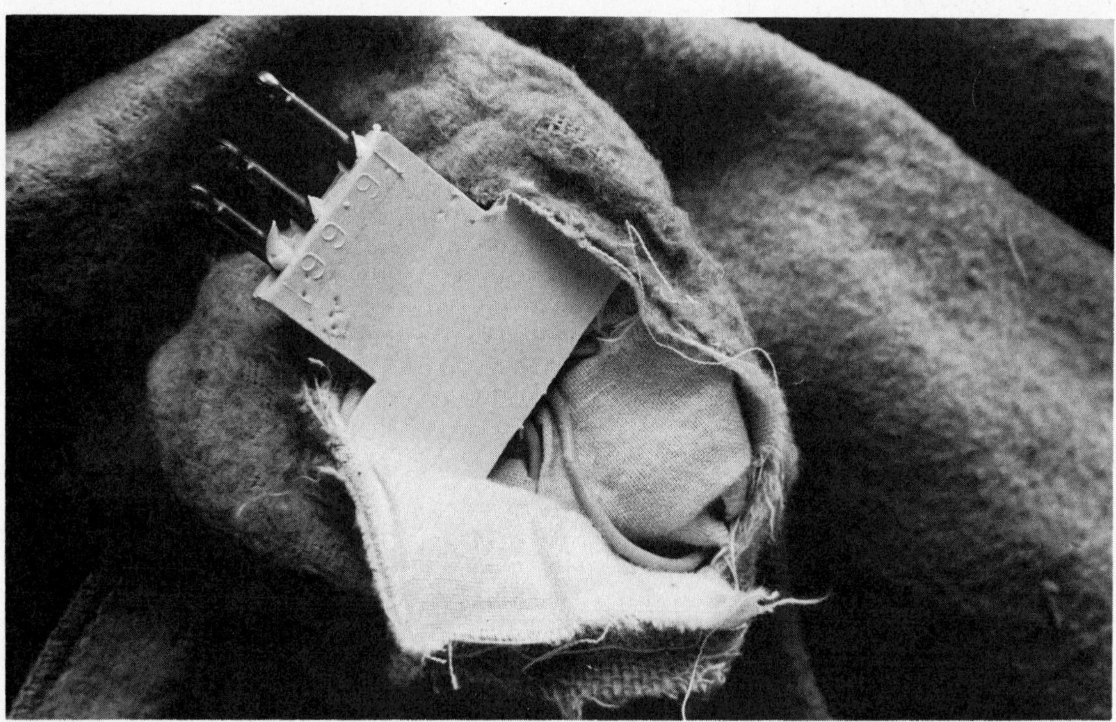

If the connector on a heating blanket becomes ripped or frayed, it is time to buy a new blanket.

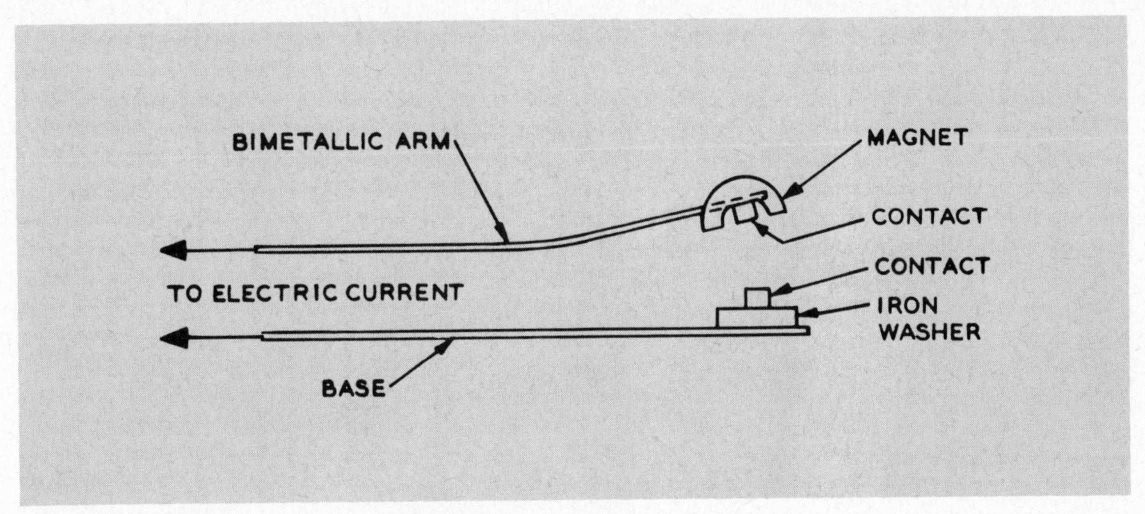

Fig. 5-3. Thermostat.

there are two separate controls on the control box, one for each heating element. In some models, two separate control boxes, each with its own thermostat, are used with a dual-control blanket.

The temperature is controlled automatically by the thermostat in the control box. This thermostat has a magnetic contact, as shown in Figure 5-3. The lower contact is mounted on the base of the control box and is surrounded by an iron washer. The upper contact is in the center of a cup-shaped magnet, and both are attached to the bimetallic arm of the thermostat. When the temperature in the room falls below a predetermined value, the thermostat arm straightens, allowing the magnet to come close to the iron washer. When it is close enough, magnetic attraction draws it the rest of the way, joining the contacts. Without the magnet, the contact points would bounce and not make a stable contact. When the temperature increases, the arms tend to separate but are held together by the magnets until the force of the bending thermostat arm overcomes the magnetic attraction. As a result electrical contact is made firmly and broken cleanly.

The original spacing between the two arms of the thermostat is adjusted by a knob on the cover of the control box. This sets the operating temperature. In addition, most boxes contain an on-off switch and a pilot light, which is usually a neon bulb.

Most troubles with electric blankets occur in the control box or line cord. The knob can be pulled off and the box opened. In some cases, the knob is held in place by a setscrew which must be loosened. Once you have the box open, you should make a visual check for loose connections or broken wires. Then make voltage tests to make sure the voltage is getting to the box, and with the plug pulled out make a continuity test to make sure the heating elements are intact. Push down the magnet to check if it is drawn to the iron washer. After many contacts, a magnet can lose its magnetism. If so, replace it with a new magnet. Or if necessary replace the whole thermostat.

If a heating element opens, it cannot be fixed. Half of the blanket will still work if the unit is a two-element electric blanket. However, even if no heat is produced don't discard the blanket, since it can still be used as a regular blanket without electricity.

5-3. Hot Plates

The simplest electrical appliance is a hot plate, consisting of a line cord and a heating element. The line cord may be permanently connected to the hot plate or may be a separate cord set. The hot plate is used to heat water in a pan or to do simple cooking where no stove is available. The heating element is usually made of coiled nichrome wire, but Calrod elements are sometimes used.

If a simple hot plate fails to operate, the trouble is either in the line cord or the heating element. This can be pinned down quickly by making one voltage check, as illustrated in Figure 5-4. The line cord is plugged into a live socket and the voltage is checked at points A and B in the figure. These are the connections between the line cord and the heating element. The probes of your test-light or meter are held in contact with points A and B. If a voltage is present, the lamp lights or the meter indicates the voltage, and therefore the line cord is not at fault. The trouble must be in the heater. With no voltage, the trouble is in the line cord. A corroborating test if the trouble seems to be in the heating element is a continuity check. With the plug pulled out of the wall outlet, check the continuity between points A and B in Figure 5-4. If the heater is open, there will be no continuity. In the case of an open nichrome coil heating element, you should be able to spot the broken wire.

Once the source of trouble is known, fixing the appliance is simple. If the cord is at fault, correct it as described in Chapter 3. If the heating element is bad, replace it with an identical element if possible.

To operate the simple hot plate, you plug it in to heat, pull out the plug to shut it off. A slightly more sophisticated model includes a switch in addition to the heating element and line cord. This unit can be left plugged in all the time and is turned on by flipping the switch. The switch is an additional source of trouble and should be checked if the hot plate fails to operate. With the plug pulled out of the receptacle, check continuity across the switch both in the off and on positions. The switch should be open in the off position and closed (continuous) in the on position. If the switch seems at fault, clip a short clip lead across it, thus removing it from the circuit. Now when you put the plug back in the wall outlet, the hot plate should begin to heat. This is another confirming test. Do not leave the clip lead in place. Replace the switch.

Hot plates are also made with two or more heating elements and a separate switch for each. In effect, such a multiple-burner hot plate is really several separate appliances in one package. If all burners fail to work, the trouble is in the line cord or line voltage (check the fuse). If only one fails, the trouble is in that burner or its associated switch, but the line cord is not at fault.

In summary, the possible faults in a hot plate lie in the line cord, switch or heating element, or their interconnections. If the hot plate blows a fuse when it is plugged in, the trouble is a short circuit, probably in the line cord. If the hot plate fails to heat, the trouble is an open circuit or a faulty switch. In any case, voltage checks and continuity checks will enable you to locate the trouble.

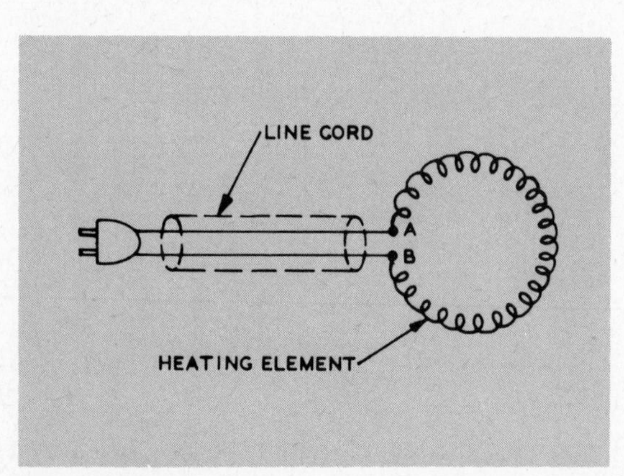

Fig. 5-4. Voltage check-points.

5-4. Popcorn Poppers

A popcorn popper consists of a covered pan, which may be made of metal or glass, supported a suitable distance above a hot plate. Oil and corn are placed in the pan. The heat raises the temperature of the oil and causes the corn to pop. If the pan were too close to the source of heat, the corn might burn before the oil reached the proper popping temperature. The pan is usually supported by a flange around the hot plate as shown in Figure 5-5.

Electric popcorn poppers are composed basically of a simple hot plate and a covered pan of metal or glass.

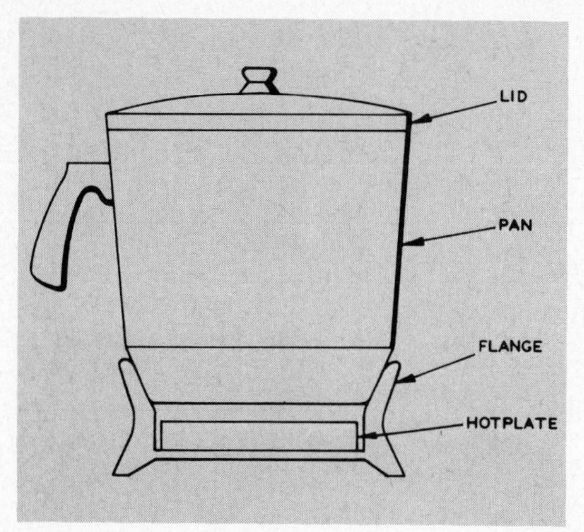

Fig. 5-5. Popcorn popper.

The electrical portion of the popper is nothing more than a simple hot plate. The possible troubles and their corrective measures are discussed in the preceding section. In some models, referred to as *automatic* poppers, a thermostat is added to shut off the popper when the temperature gets too high, which usually happens after the corn is popped. The thermostat should be treated simply as a switch. It should be closed when it is cold so that the heating element begins to heat immediately when the popper is plugged into an outlet.

One possible source of trouble is a thermostat which opens at too low a temperature so that the oil never gets hot enough to pop the corn. If the oil in an automatic popper gets warm, but nothing else happens, the trouble is probably the thermostat opening too soon. Try bypassing the thermostat. If this corrects the trouble, you can leave it that way, since it will operate quite well as a corn popper, although it will no longer be automatic. When you get around to it, you can replace the thermostat and make the popper automatic again. If bypassing the thermostat does not cure the trouble, the fault is probably a broken wire in the heating element. When cold, the two ends of wire touch, completing the circuit, and the element gets hot. However, as the heater gets hot, it expands, opening the break. In effect, the broken heater is acting like a thermostat. If this is happening, you should be able to spot the break when the heater is getting hot. To fix it, replace the heater element.

A simple hot plate, with or without a thermostat, is also the only electrical part in a few other electrical appliances, notably old-fashioned vacuum coffee pots and deep-fat fryers. The possible faults and corrective measures for these appliances are essentially the same as for the simple hot plate described in this and the preceding sections. For the most part, manufacturers have done away with the separate hot plate; modern appliances have the heating elements built into them.

5-5. Waffle Irons

The electrical portion of a waffle iron consists of a line cord, switch, thermostat, two heating elements and pilot light. The two heating elements, one in each half of the iron, are wired in parallel so that they are turned on and off together. The pilot light may be a small bulb, a neon light, or in some inexpensive models, simply a small window on the outside of the case. When a window is used, the glow from a heating element is visible and serves as a pilot light.

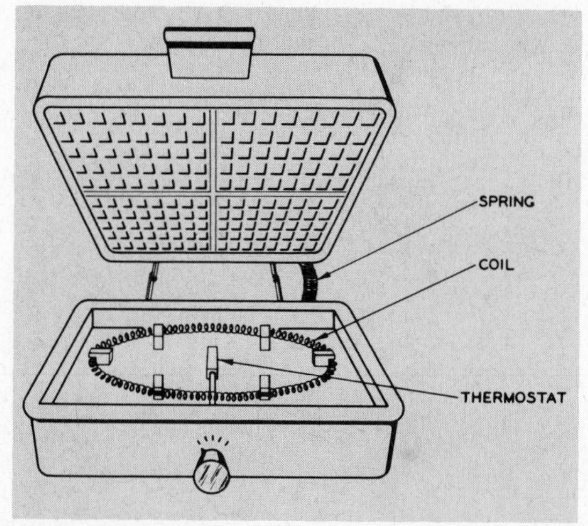

Fig. 5-6. Waffle iron.

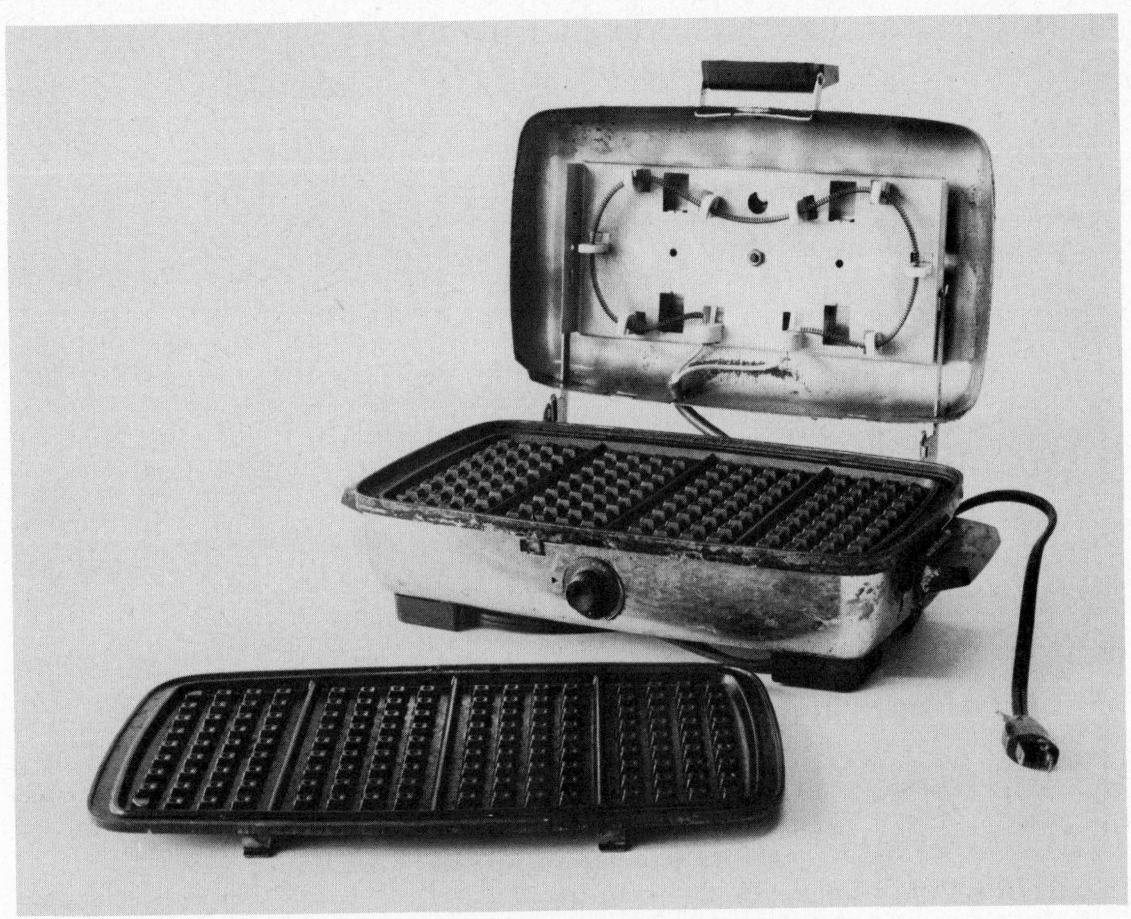

The heating element in this waffle iron is a coiled nichrome wire supported on porcelain insulators.

The heating elements in a waffle iron are usually coiled nichrome wire supported on porcelain insulators. Each element is mounted in a removable metal pan. The two elements are connected by wires passing through a protective spring on the hinge side of the iron. A waffle iron with bottom grill removed is shown in Figure 5-6. Note the coil spring joining the two halves of the iron. Without this spring, the wires joining the two heating elements would be subject to too much abuse with the frequent opening and closing of the iron. The coiled nichrome element is also visible in the photograph, as is the thermostat in the center of the bottom pan. The line cord is usually connected directly to the lower heating element, and the voltage is brought to the upper element by means of two insulated wires that pass through the protective spring.

Servicing a faulty waffle iron involves locating the source of trouble, figuring out how to get at the defective part, and repairing or replacing the part. After the repair the iron should be tested to make sure that it operates properly and that *it is safe to operate.*

If the iron does not heat at all when plugged into a live outlet, the trouble may be due to a defective line cord, a loose connection, a defective thermostat which remains open, a faulty switch, or (most unlikely) both heating elements failing simultaneously. Remove the grids. Loosen the screws that hold the pans in place, and lift the pans holding the heating elements. Using voltage and continuity tests, check the line cord and thermostat. Look for loose connections. Unless the waffle iron was dropped or otherwise shaken up, the trouble is probably in the thermostat or line cord. If it proves to be the line cord, you can fix it by using the techniques described in Chapter 3. If the thermostat is at fault, replace it. It is usually held in place by screws. The control knob on the outside is held on by friction and can be pulled off. When the new thermostat is installed, replace the knob so that it turns over the whole scale on the dial. If the switch is bad, replace it.

If one side heats, but the other does not, the trouble is probably a heating element. Check the element which doesn't heat for continuity, and if it is open, replace it. If the upper element doesn't work, the trouble may also be in the wires passing through the spring. Test these for continuity. If one is bad, replace both, since it is easier, and the second wire may be worn and ready to break. When replacing these wires, slip them through the spring together; you should allow four or five inches of slack in the leads to permit flexing without strain.

If the iron doesn't get hot enough, the trouble is probably due to the thermostat opening too soon. Check the line voltage first in this case, since a low line voltage can cause this trouble. If the fault is in the thermostat, replace it.

If the iron gets too hot, the trouble is almost certainly caused by a thermostat that won't open. Replace it.

If the pilot light fails, but the waffle iron works perfectly otherwise, you may ignore the trouble. If you decide to replace the bulb, make sure the new one has the same wattage as the one removed.

Broken feet or handles do not affect the electrical operation, but can be annoying. To get at the screws holding them, it is usually necessary to remove the grids and pans holding the elements. The broken parts can then be replaced. Note carefully the positions of all elements so that they can be replaced correctly.

When you have finished repairing a waffle iron, make a ground check to make sure there is no voltage on the case of the iron. Do this with the plug pushed into the outlet in both positions so that first one and then the other side of the line cord is grounded.

5-6. Sandwich Grills

Sandwich grills and special grills to make party snacks are electrically identical to waffle irons and are repaired in the same manner. The only problem may be taking the grill apart. The method of construction varies, but look for screws at the bottoms of deep holes in handles or brackets if the fastenings are not otherwise obvious. Always make a ground check when you finish repairing the grill.

5-7. Electric Skillets

An electric skillet or fry-pan has a built-in heating element which is not removable. This heating element should not burn out in normal operation and is unlikely to be damaged unless the skillet is dropped with sufficient force to break the connections to the electric contacts. If the heating element is damaged, it is not repairable, and you may as well get a new skillet. The heating element will not be damaged if the skillet is immersed in water, so washing it after use is no problem.

The parts of an electric skillet are shown in Figure 5-7. The heating element is fastened to the bottom of an aluminum pan. An aluminum sleeve is fastened to the pan in such a manner that the sensing probe on the control unit can be plugged into it. The pan also has a Bakelite handle which is usually removable. The skillet is usually enclosed in a decorative metal housing which encloses the heating element and has a cover to match.

The control unit contains the thermostat. When the control unit is plugged into the receptacle on the skillet and turned on, the line voltage is connected across the heating element. The probe is inserted in the sleeve between the ends of heating element. As the skillet gets hot, the sleeve also gets hot and transfers the heat to the probe. The thermostat is fastened directly to the probe and, at the proper temperature, opens to shut off the electricity. The temperature of operation is

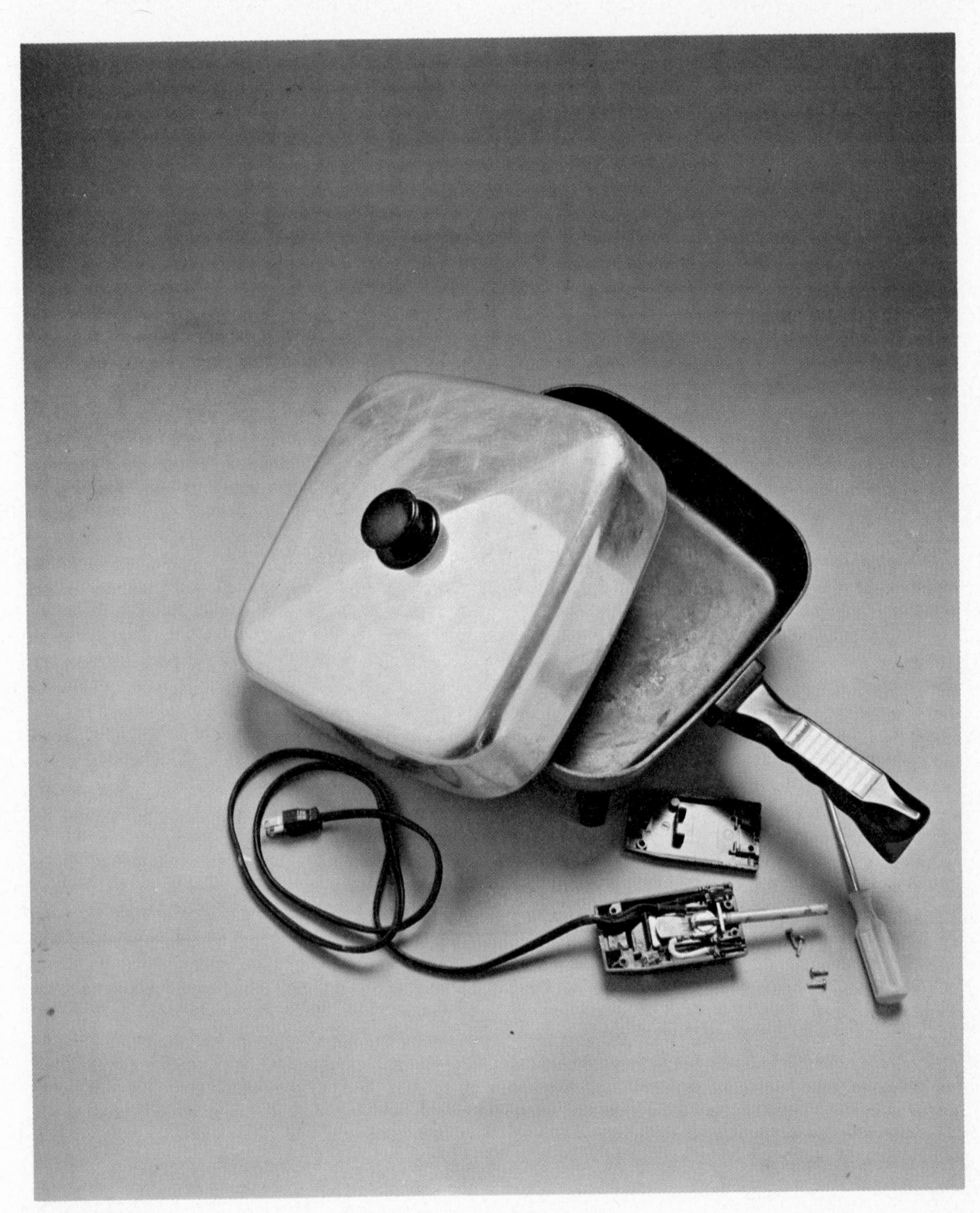

Electric skillet.

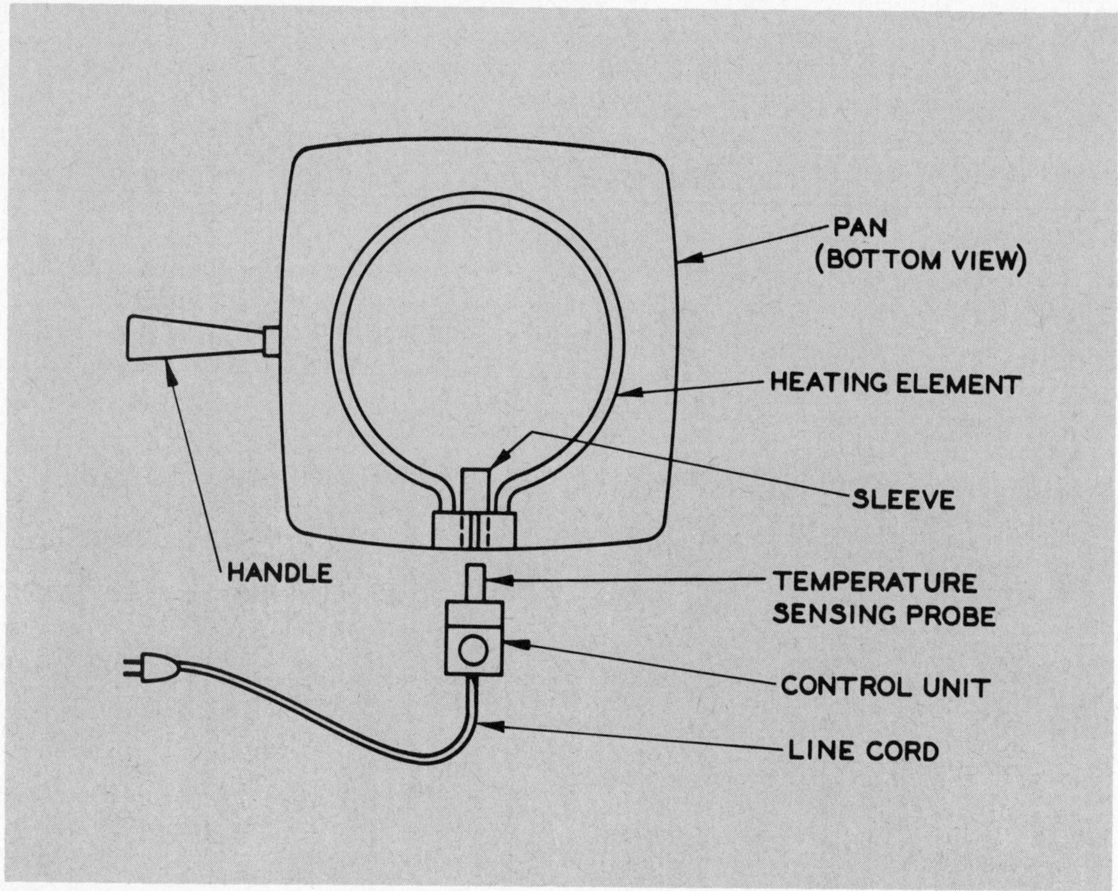

Fig. 5-7. Electric skillet.

adjusted by a small knob which also acts as the on-off switch. A pilot light on the control unit indicates when the skillet is drawing current. *The control unit must never be immersed in water.*

If the skillet does not heat when connected to a live outlet, the trouble may be in the line cord, heating element, switch or thermostat. As a first check, plug in the line cord, but do not plug the control unit into the skillet. Check the voltage at the output of the control unit by inserting your meter probes into the contacts there. Make sure the switch is turned to the *on* position. If you find voltage at the output of the control box, the trouble is probably the heating element. This can be verified by making a continuity check on the heating element. Note that if both are good, the trouble is in the contacts and these should be cleaned. If the heating element is open, get a new skillet. It is not necessary to get a new control box.

If there is no voltage at the output of the control box, then you can take the box apart and check the line cord, switch and thermostat separately. Line cord repairs are discussed in Chapter 3. Replace a defective switch or thermostat with an identical replacement part.

If the skillet heats to the wrong temperature, the thermostat may be dirty, corroded or out of adjustment. Clean the sensing element and all contacts in the control box. If necessary, replace the thermostat. After finishing the repairs always check the case of the skillet for voltage as a safety measure. Do this with the plug in both positions in the outlet.

When washing an electric skillet, be sure to

Never immerse the control unit of an electric skillet in water.

Percolator

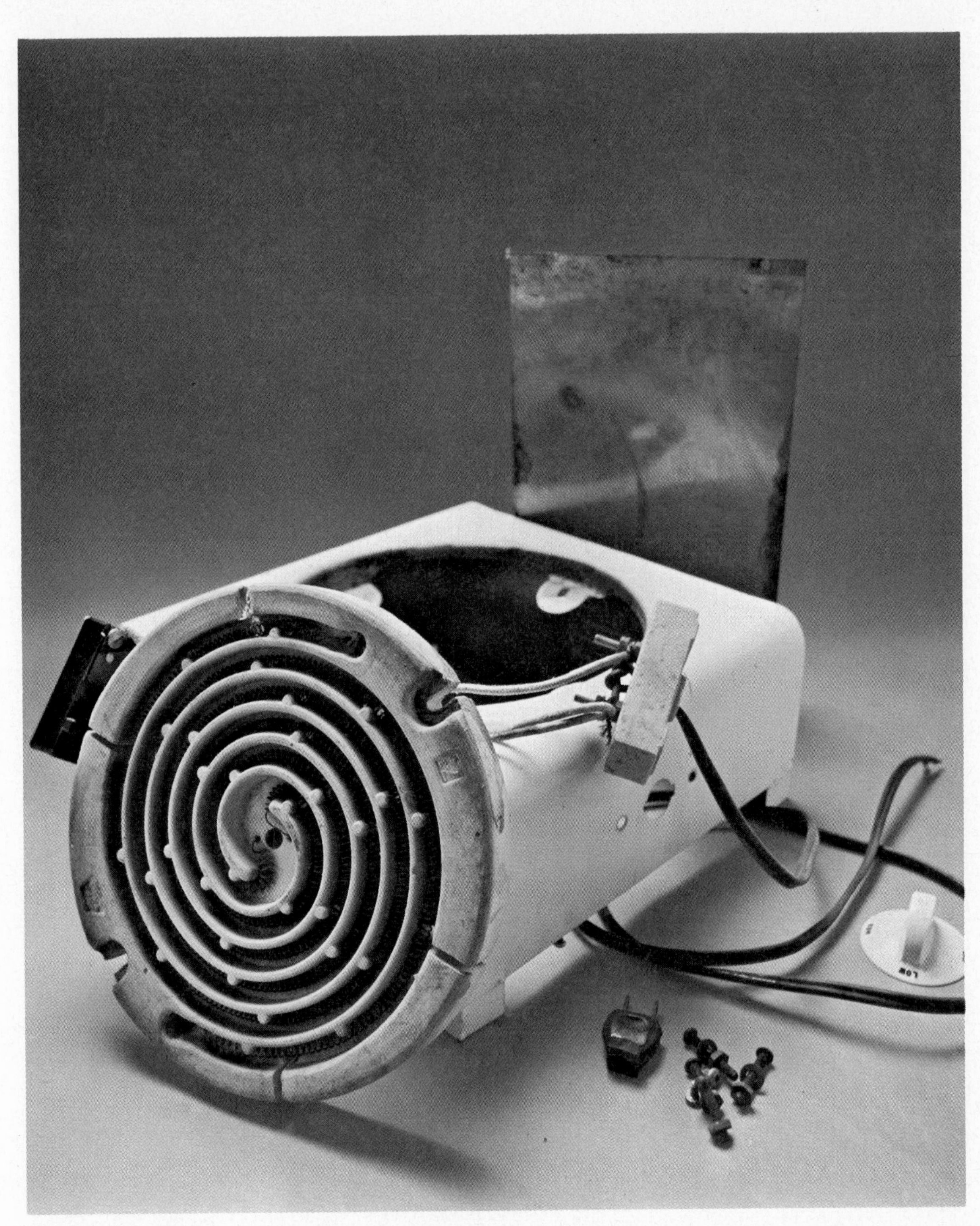

Interior of hot plate.

follow the manufacturer's instructions. Disconnect the control box before washing. Do not use harsh or abrasive cleaners when cleansing the unit. Wash the pan as soon as possible after use. Don't soak any part in water and dry everything promptly after washing. Don't store in an oven since the plastic parts can be damaged by heat. If these rules are not adhered to, the surface of the pan may discolor. This may be esthetically displeasing, but discoloration does not interfere with proper operation of the skillet, and it has no effect on the food.

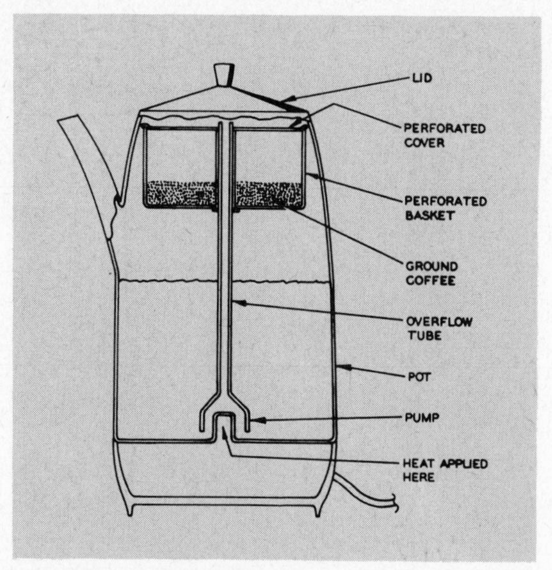

Fig. 5-8. Percolator action.

5-8. Percolators

Electric coffee makers or percolators are of two types, non-automatic and automatic. The brewing action of both types is the same and is shown in Figure 5-8. Cold water is placed in the pot and ground coffee in the basket at the top of the overflow tube. The bottom of the overflow tube contains a valve which shuts when pressure builds up inside the tube. This part of the percolator is called the *pump*. Heat is applied to the cold water inside the pump causing it to boil. The pressure closes the valve, and the steam pushes the hot water up the tube, where it is ejected from the top. The water hits a deflector in the lid and falls back onto the basket cover. The perforations in the cover and basket permit the water to ooze through the coffee grounds slowly. After the water has been pushed out of the tube, the pressure inside decreases, and the valve opens allowing more water to enter the pump. The process is then repeated. The electrical portion of the percolator is usually enclosed in a Bakelite housing at the bottom of the appliance. In some types the heater surrounds a well in the bottom of the pot, and the overflow tube and pump are inserted in this well. The principle of operation is the same.

Non-automatic percolators do not shut off automatically. When the user thinks the coffee is done, he pulls the plug. To reheat the coffee or keep it warm, he first removes the basket and tube and then plugs the unit in again. The electrical portion of the pot is thus very simple. It consists of a line cord, a heating element and a fuse. The fuse prevents the heating element from burning out, if the percolator is left plugged in without any water in it.

If a non-automatic percolator won't heat, the trouble must be in the fuse, the line cord or the heating element. The Bakelite housing can be removed by loosening a screw or a nut in the center of the bottom of the base, thus exposing the circuitry. The fuse should be checked first since it is the part most likely to fail. Using continuity tests you can determine which part is at fault. For a bad line cord, follow the practices discussed in Chapter 3. If the fuse or heating element has failed, replace it with an identical part. Make sure no part of the electric circuit makes contact with the metal pot before you return the percolator to service.

Automatic percolators shut off by themselves when the coffee is brewed, and then heat is applied to keep the coffee warm without percolating further. Although there are almost as many different circuits as there are manufacturers, all are thermostatically controlled, and the principles of operation are similar. Some have a pilot light which goes on

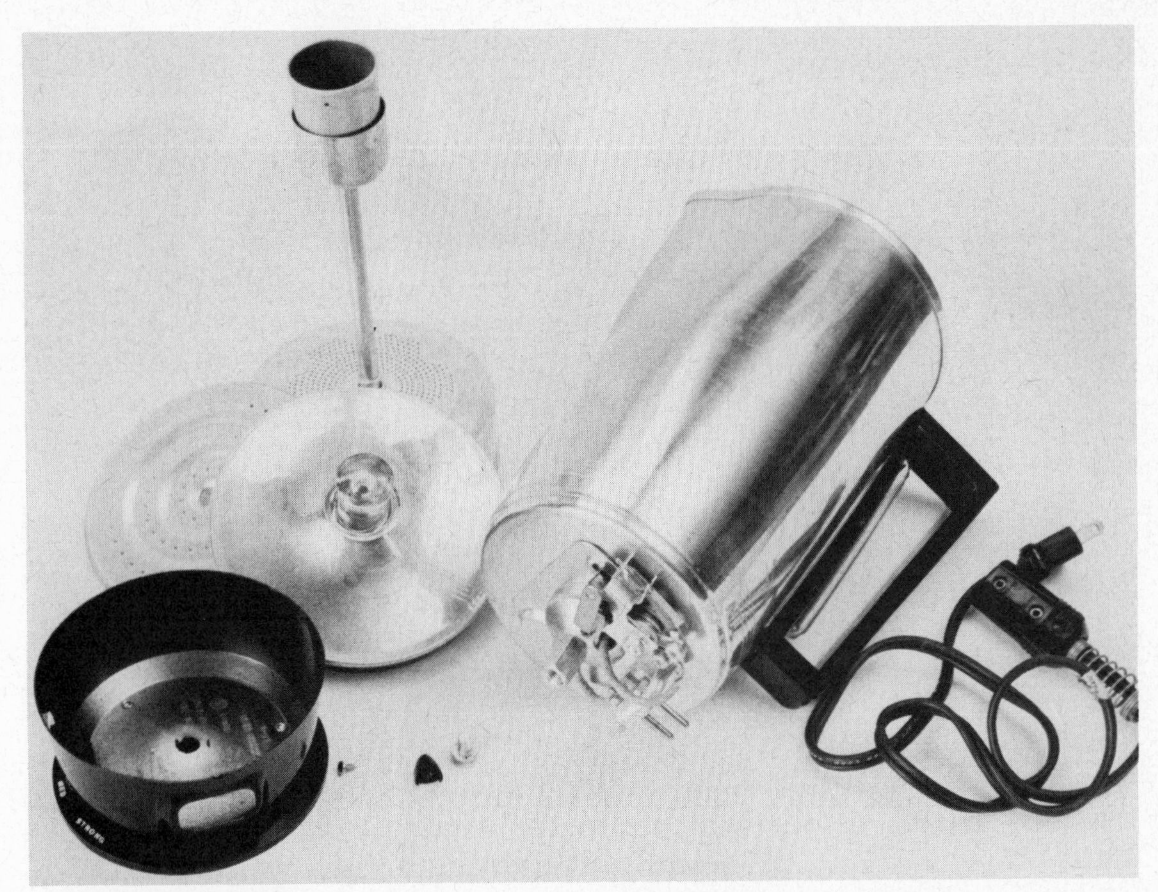

The electrical part of most percolators is enclosed in a Bakelite housing at the bottom of the appliance.

when the coffee is brewed. A typical circuit is shown in Figure 5-9.

During the brewing cycle, the thermostat in Figure 5-9 is closed, shorting out the signal light and the warming element. The full line voltage is across the main heating element,

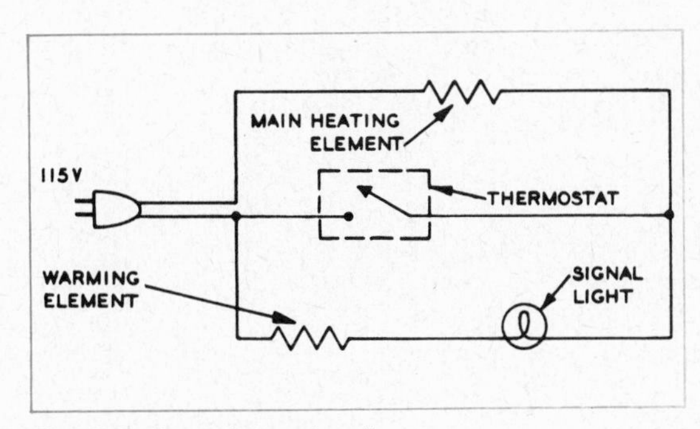

Fig. 5-9. Percolator circuit.

which heats the water in the pump, causing the percolating action. Although the water inside the pump is boiling, the rest of the water in the pot may still be cool. However, the hot water drips through the coffee grounds and falls back in the pot as coffee, and more cold water enters the pump to be heated. As a result the coffee in the pot gets stronger and hotter. The longer the percolation, the stronger the coffee will be.

When the brewed coffee reaches a predetermined temperature, the thermostat opens. Now the main heating element, the warming element, and the pilot light are in series across the line voltage. The portion of this voltage across the main heating element is small, so that this element no longer gets hot enough to boil the water in the pump. Most of the voltage is across the warming element

Modern automatic percolator.

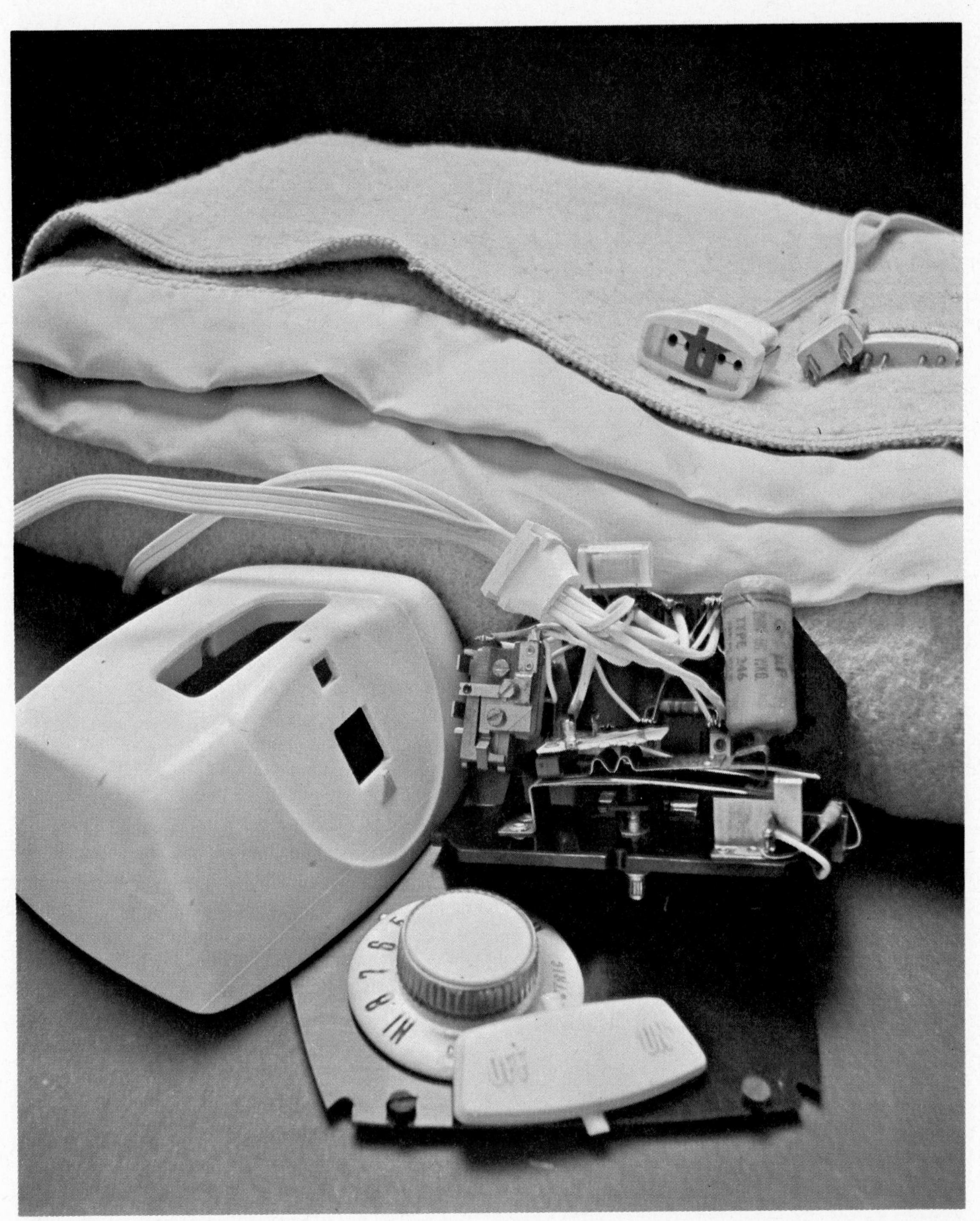

Control box of electric blanket.

which warms the bottom of the pot outside of the pump. The pilot light indicates that brewing has ceased, and the warming element keeps the coffee warm. On some models, the temperature at which the thermostat opens can be adjusted by a control on the outside.

Setting the control for stronger coffee simply means raising the temperature at which the thermostat opens, thereby prolonging the brewing action. The same effect can be achieved by starting with colder water. Generally, the thermostat should open at about 190 degrees when the control is set for strong coffee.

As with non-automatic models, the base of an automatic percolator is a Bakelite housing which is held on by a screw or nut in the center. If there is a control to adjust the brewing time, this should be removed before removing the base. Loosen a setscrew on the control knob, if there is one; if not, simply pull it off.

If an automatic percolator won't heat at all, the trouble is probably in the line cord or the main heating element. Note that if the warming element failed, the main heating element would still operate. If the thermostat failed in the open position, both heating elements would work, and the water would get warm, although it wouldn't percolate. If the thermostat failed in the closed position, percolation would take place but it would never stop. If the trouble is in the line cord, fix it, referring to Chapter 3. If the main heating element is burnt out, replace it with an identical part. Continuity tests should enable you to determine which part is defective.

Notice that if the warming element or the pilot lamp fails, the percolator will work until the coffee is brewed, but then the pilot lamp will not go on, and the coffee will not be kept warm. Check both the lamp and the warming element for continuity, and replace the defective part. On models without a pilot lamp, if the coffee cools off after brewing, it is due to a defective warming element.

There may also be mechanical failures. If the valve is bent or broken, the percolator will not brew even though the electric circuit is faultless. A bent valve is easily replaced. Other mechanical defects include broken Bakelite base, leaky gaskets, broken handles and cracked lids. All these parts are obtainable from an authorized service dealer or the manufacturer and are easily replaced. Figure 5-10 is an exploded view of the Toastmaster Model M521 Coffee Maker manufactured by the Toastmaster Division of McGraw-Edison Company. Table 5-1 lists the parts in Figure 5-10. Notice particularly parts number 3, 8, 10 and 15. These are gaskets and washers to prevent leaks. Whenever a mechanical defect requires replacement of a part, you must remember to put in a new sealing device also, since the old one may be deformed during removal of the defective part.

As with all appliances, always check for voltage on the case as a safety measure before returning the percolator to service. Make ground checks with the plug in both positions in the outlet.

Table 5-1. Toastmaster Stainless Steel Coffee Maker – Model M521.

No.	Part
1	top cover knob, black
2	top cover, stainless steel
3	teflon washer (for cover knob screw)
4	machine screw 8-18 x 3/8 (top cover knob)
5	basket cover, stainless steel
6	basket and guide tube assembly, stainless steel
7	percolator tube and pump assembly
8	spout gasket
9	spout mounting screw
10	teflon washer (handle screw)
11	handle screw
12	handle, black
13	sight tube "see level"
14	sight tube reflector, aluminum
15	sight tube seal, rubber
16	body w/percolator element and keep warm element bracket

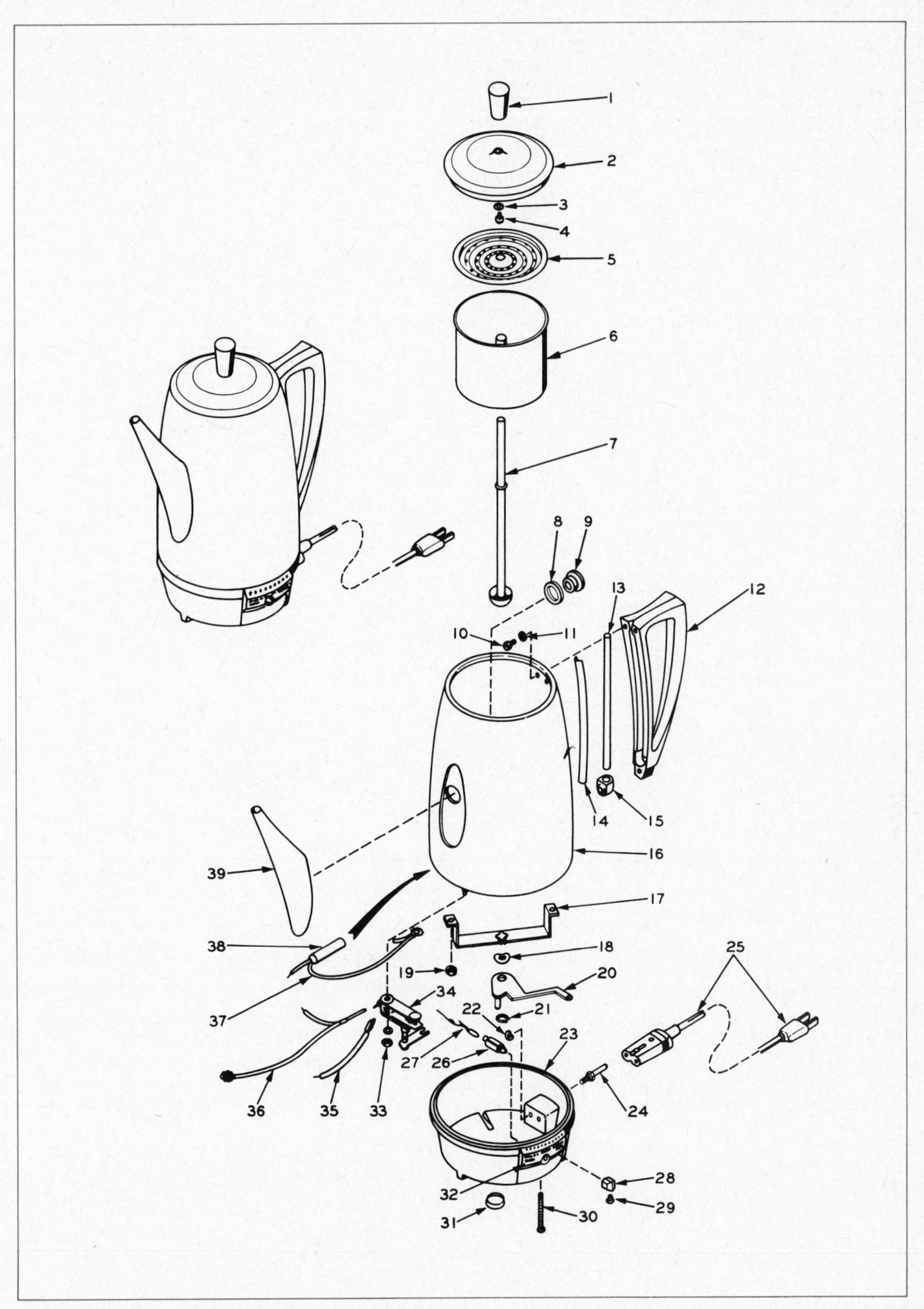

Fig. 5-10. Toastmaster model M521 coffee maker.

17 base bracket and stud assembly
18 pivot tension spring
19 nut base bracket, # 10-32 x 3/32 x 1/4 hex brass
20 flavor control arm and stud assembly
21 truarc retaining ring
22 hex nut, (terminal pin) 6/32 x 3/32 x 1/4 steel
23 Bakelite base, black
24 terminal pin
25 cord and plug, black
26 signal light lens
27 signal light and resistor assembly
28 flavor control knob
29 flavor control knob screw 3/48 x 1/8 fil. h.ss
30 Bakelite base screw 5/32 x 1/2
31 hole plug 17/32" diameter
32 name plate (on base)
33 nut 8/32" x 1/8 x 11/32" (holds thermostat in place)
34 thermostat assembly
35 3" lead wire, thermostat to percolator element
36 7" lead wire, terminal to thermostat
37 3-5/8" lead wire, percolator element to terminal
38 keep warm element cartridge
39 spout and insert assembly

5-9. Electric Irons

There are two types of electric iron: dry and steam. Electrically they are identical, each consisting simply of a line cord, a thermostat and a heating element. The steam iron has a water chamber in addition to the electrical and mechanical parts of the dry iron. The external parts of an iron are illustrated in Figure 5-11.

There is a variety of shapes and sizes of

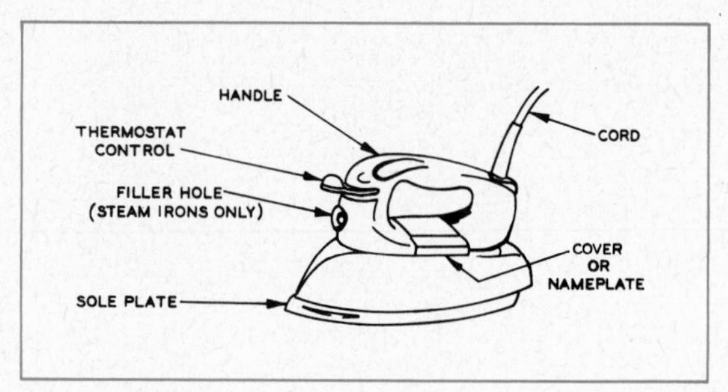

Fig. 5-11. Iron.

heating elements in irons. Nichrome ribbon, nichrome coils and Calrod are all used. Nichrome ribbon is usually wound on mica sheets and is replaceable if it burns out. The other two types are built into the sole plate, but electrically insulated from it, and cannot be removed. Practically all modern irons use non-replaceable heating elements. This means that when the heating element burns out, the iron should be discarded. Fortunately, with proper usage the heating element will outlast the other electrical parts of the iron.

In Figure 5-11, a thermostat control is shown on the handle and is located so that it may be moved by the finger tips when holding the iron normally. This control is connected by a rod to the thermostat near the sole plate, and it regulates the opening between the contacts and thus the temperature at which the thermostat opens and closes. The thermostat itself comes in many shapes, but the principle of all those used in irons is based on the bending of a bimetallic arm.

When working on any electric iron, always make sure your work bench is covered with a soft pad or several layers of cloth. If the bottom of the sole plate is nicked or scratched, the iron will be useless, even if the electrical parts work perfectly.

The line cord and its connections to the rest of the circuit are the chief sources of trouble in an iron. This is not surprising, since the cord is continually being flexed when the iron is in use. Thermostats sometimes burn out or stick, but thermostat troubles are rarer than mechanical problems caused by dropping the iron.

Fixing an iron is very simple once you have

An electric iron consists basically of a line cord, a thermostat and a heating element.

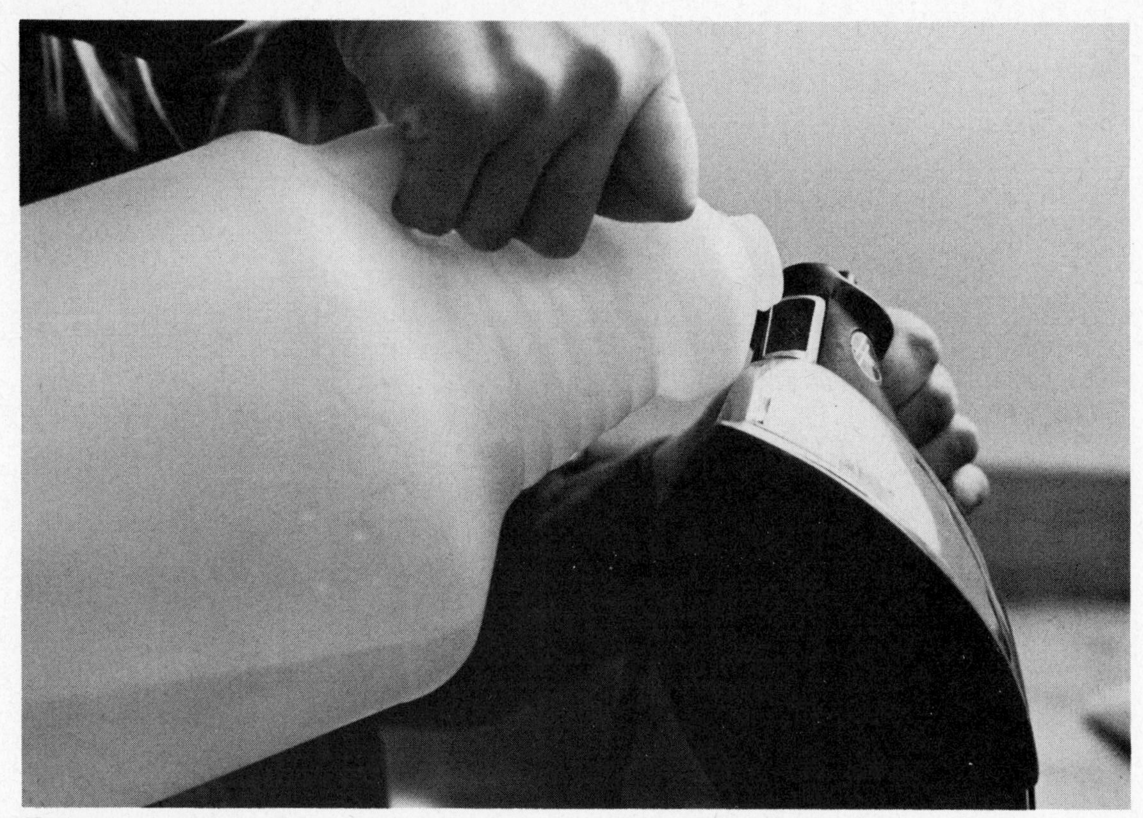

To clean out mineral deposits that build up in an iron, fill it with vinegar and let it soak for 12 hours.

figured out how to take it apart. Each model is a puzzle, but once the key is found, the iron can be disassembled easily. Never try to force it apart. On many models, a flexible metal cover, sometimes used as a nameplate, conceals a large nut in the center of the iron. This cover is shown in the illustration of the iron. It can be lifted off, giving access to the electrical circuit. The temperature scale may have to be pried forward and off on some models before the handle can be lifted from the rest of the iron. On steam irons, the filler tube should be pulled forward and out before lifting off the handle. In order to be sure that you reassemble the iron with the thermostat control in the same position as when you disassembled it, you should push this control to one extreme end before disassembling. Then, when you reassemble the iron, make sure the control is at the same end.

The cord is usually attached to the rest of the circuit by two spade lugs which are clamped under their respective screw heads. To remove the cord, loosen the screws, and the cord can be pulled out. It is not necessary to remove these screws completely.

At least one model is held together by a plastic pin located at the top of the handle near the front. To take this iron apart, drive this pin into the iron until the handle is loose. The pin can be recovered after the iron is taken apart and should be replaced in its original position when the iron is reassembled.

In some models, it is necessary to pry off the indicator arrow at the front of the handle and to remove the control knob and adjusting screw. This will reveal a square hole at the front of the handle. You should insert a Phillips head screwdriver into this hole and loosen the screw at the bottom, which holds the iron together.

After you have the iron apart, it is a simple matter to make voltage tests and continuity tests to determine the trouble. With the plug inserted in a live receptable, you should find voltage across the terminal screws of the line cord. Check this voltage if you have a meter, since an iron works poorly if the line voltage drops as little as 10 per cent. However, in most locations, a line voltage drop is rare.

If the iron won't heat, the trouble is in the line cord, heating element or thermostat. There is nothing else. If there is no voltage at the line cord terminals inside the iron, the line cord is a fault. Fix it, using the techniques discussed in Chapter 3. If the line cord is all right, make continuity checks on the thermostat and heating element. Usually the thermostat can be replaced if it is bad, but the heating element cannot be replaced. Note carefully how the thermostat is connected, and if it needs replacement, use an identical part.

If the iron gets warm but not hot enough to iron, the trouble may be a low line voltage (possible, but unlikely) or a thermostat that is loose or shuts off too soon. Put a clip lead across the thermostat, bypassing it. If the iron now gets hot, the trouble is in the thermostat. The contacts may be dirty or bent. Generally, it is best to replace the unit rather than try to fix it.

A steam iron is identical to a dry iron electrically, but may succumb to troubles caused in the water chamber. All manufacturers recommend using distilled water in their steam irons, but some users forget or need to iron when no distilled water is available. If tap water is used, impurities cause deposits inside the water reservoir in the iron, and in the valve and escape holes. When these become clogged, the iron can still be used as a dry iron, but not as a steam iron. To clean out the deposits, soak the unit for at least 12 hours in vinegar. The deposits can then be flushed out with water. A steam iron may also develop leaks. Take the iron apart and replace any gaskets or sealing washers which show signs of deterioration.

Before reassembling an iron make sure that no stray strands or wire will contact the case of the iron when it is put back together. After assembly check the iron for voltage on the case. This ground check should be performed for both positions of the plug in the receptacle.

5-10. Toasters

Although there are many different types of electric toasters on the market, to most people an electric toaster is the popular pop-up model. Slices of bread are dropped into slots in the top of the toaster and are moved down into the body of the appliance either manually or automatically. When the bread is toasted, it pops up, and the toaster shuts itself off. Other types, resembling sandwich grills, table ovens and the like, are relatively simple devices and are repaired by the techniques listed under waffle irons or hot plates. The automatic pop-up toaster is *not* a simple device to repair. Both the electrical circuit and the mechanical operation are more complicated than in most other small electrical appliances.

To make matters more complicated, there are many different kinds of pop-up toasters operating on widely different principles. If you remove the case from a toaster while it is still in operating condition, you might be able to figure out its method of operation by watching it work. But if it is out of order, this becomes a difficult task. Fortunately, all toasters give many years of trouble-free service, and when they do go wrong, their owners usually welcome the excuse to buy the latest improved model. Nevertheless, some repairs are relatively simple, and you can do them if you do not let yourself be awed by the apparently complicated machinery.

The first pop-up toasters were timed by a spring-wound motor. When the lever was pushed down, it wound the timing motor and turned on the switch. Then when the motor ran down, the electricity was turned off and the latch holding the carriage down was released. The carriage snapped up, bringing the toast partially out of the toaster. In order to prevent the toast from flying out completely, a pneumatic stop, similar to that used to prevent screen doors from slamming, cushioned the movement of the carriage. A color control knob on the toaster regulated the time that the carriage stayed down and the electricity remained on. At any one setting of the knob, the duration of the toasting cycle was con-

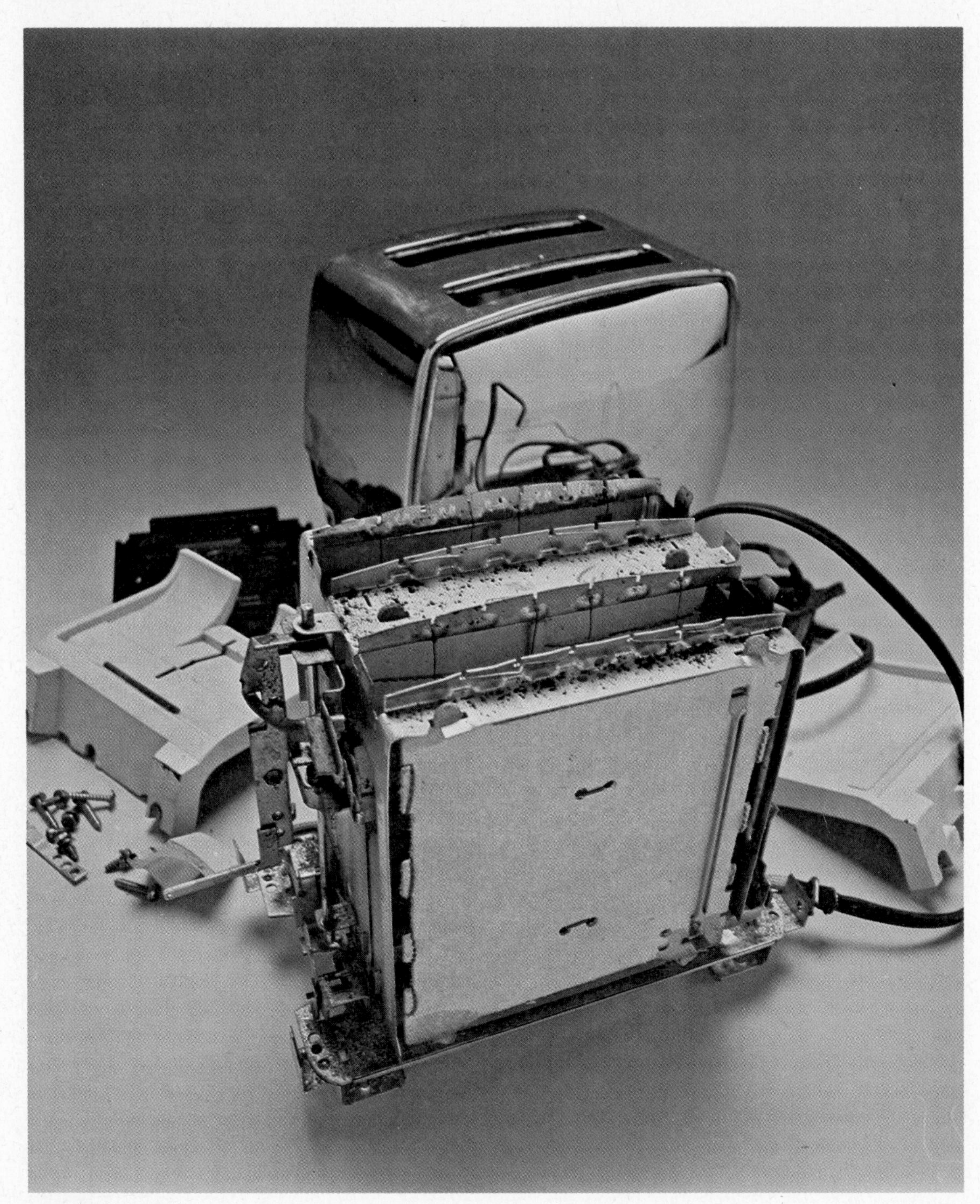

Interior of toaster.

This modern toaster may be regulated to warm pastry as well as make toast.

stant. Therefore, the first slice, put in when the toaster was cold, would not be as dark as subsequent slices of toast. The user was advised to preheat the toaster by letting it operate through one cycle before putting in the first slice.

Preheating the toaster was a slight nuisance, and naturally manufacturers devised better toasters which eliminated this problem. Although many of the old clock-timed toasters are still in use, they haven't been manufactured for years and you should not concern yourself with how to repair them. However, if you know how to fix a modern pop-up toaster, the old clock-timed model will pose no problems.

One type of automatic toaster uses a bimetallic regulator or compensator along with the clock mechanism. The regulating portion of the toaster is shown in Figure 5-12. When the toaster is cold, the bimetallic arm is straight and has no effect on the timing apparatus. As the temperature rises inside the toaster, the bimetallic arm bends toward a lever which regulates the speed of the clock. As the temperature increases further, the bimetallic arm pushes the speed-controlling lever and causes the clock to run faster, shortening the toasting time. Thus, the second and subsequent slices of toast are no darker than the first. The original position of the speed-controlling level is set by a control

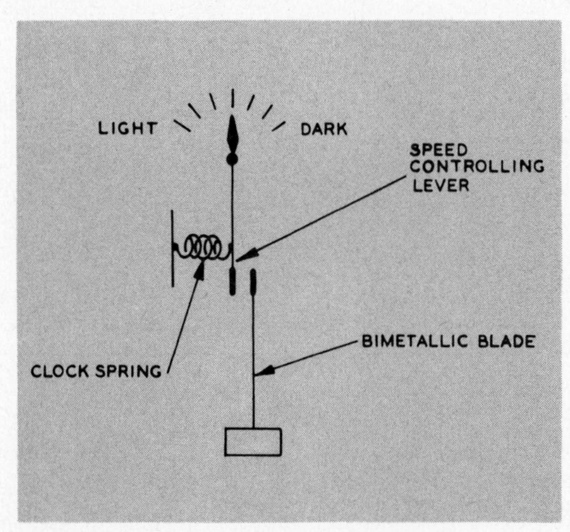

Fig. 5-12. Clock-timer with compensator.

knob on the case to choose light to dark toast.

Note that a *bimetallic arm* is used as a compensating device, but it is not called a thermostat. A thermostat also uses a bimetallic arm, but strictly speaking a thermostat is a switch which is turned off and on by heat.

In another type of automatic toaster using a similar speed-regulating control, the bimetallic arm is in contact with the bread to be toasted. If the bread is fresh, it contains more moisture and takes longer to reach a toasting temperature. The bimetallic arm remains straight for a longer period of time. On the other hand, if stale dry bread is used, it gets hot quickly, and the bimetallic arm bends earlier to speed up the clock. This type of toaster not only maintains constant color for later slices of toast, but it does so for stale and fresh bread alike.

In both types of toaster described above, when the clock runs or stops, some sort of mechanical device shuts off the electricity and moves the latch holding the carriage down. The carriage pops up, cushioned by the shock absorber, so that the toast can be removed. The toaster is then ready for the next cycle.

One type of toaster uses a *hot-wire carriage release.* A separate strand of wire is connected across the line voltage. Current flows through the wire causing it to get hot and expand. The wire holds the carriage down by means of a mechanical linkage. When the thermostat shuts off the current, the wire cools and contracts, releasing the carriage so that it pops up.

Another type of automatic toaster uses a two-stage thermostatic control. This toaster has an auxiliary heating element wrapped around the bimetallic arm. A schematic diagram of the electric circuit is shown in Figure 5-13. When the carriage is lowered, the main switch is closed, causing current to flow through both the auxiliary heater and the main heating elements. The heat from the auxiliary heater causes the bimetallic arm to bend. This is the first stage. The bent arm causes the auxiliary switch to close. Referring to Figure 5-13, you can see that this shorts out the auxiliary heater from the circuit, but current still flows through the main heating

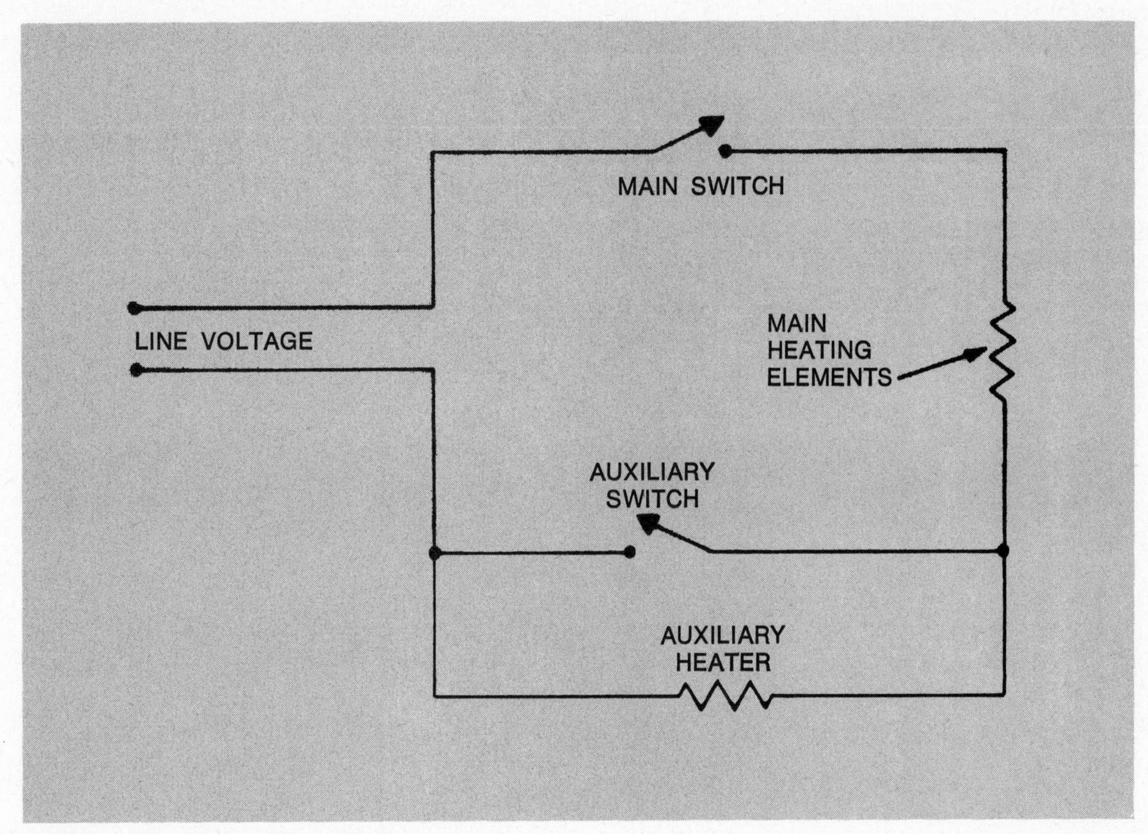

Fig. 5-13. Two-stage circuit.

elements. In this second or cooling stage, the bimetallic arm tries to straighten. However, at the same time that the arm closed the auxiliary switch, a mechanical device gripped the arm in such a way that it could not return to its original position. Then in trying to straighten, one end moves toward a release which trips the carriage, permitting it to pop up. When the carriage moves up, it opens both switches and releases the bimetallic arm, thus resetting the toaster for the next slice of bread.

The mechanical device which detains the bimetallic arm in the two-stage circuit takes many forms. One possible configuration is shown in Figure 5-14. In (a), the bimetallic arm is horizontal and straight. In (b), the arm has bent because of the heat and a mechanical detent has moved under the arm. Now when the arm cools, the detent prevents the arm from returning to its original position (shown by the dotted lines). Thus it straightens at an angle to the horizontal, and its free end trips a lever, freeing the carriage. The detent is released when the carriage moves up, and the bimetallic arm then moves back to the position shown in (a).

In the first two-stage toasters, the auxiliary heating element was an electrically insulated coil slipped over the bimetallic arm but not fastened to the arm otherwise. In later models, the heating element is firmly fixed in place, so that it is always located at the same position on the bimetallic arm.

The heating elements in all pop-up toasters are made of flat nichrome wire wound back and forth close to the positions that the bread will occupy. These heating elements are *not insulated* electrically and it is dangerous to touch them when the toaster is plugged in, even if the toaster is off. This is explained in Figure 5-15, which depicts the main circuit in the toaster. When the toaster is plugged into an outlet, one side of the heater circuit is

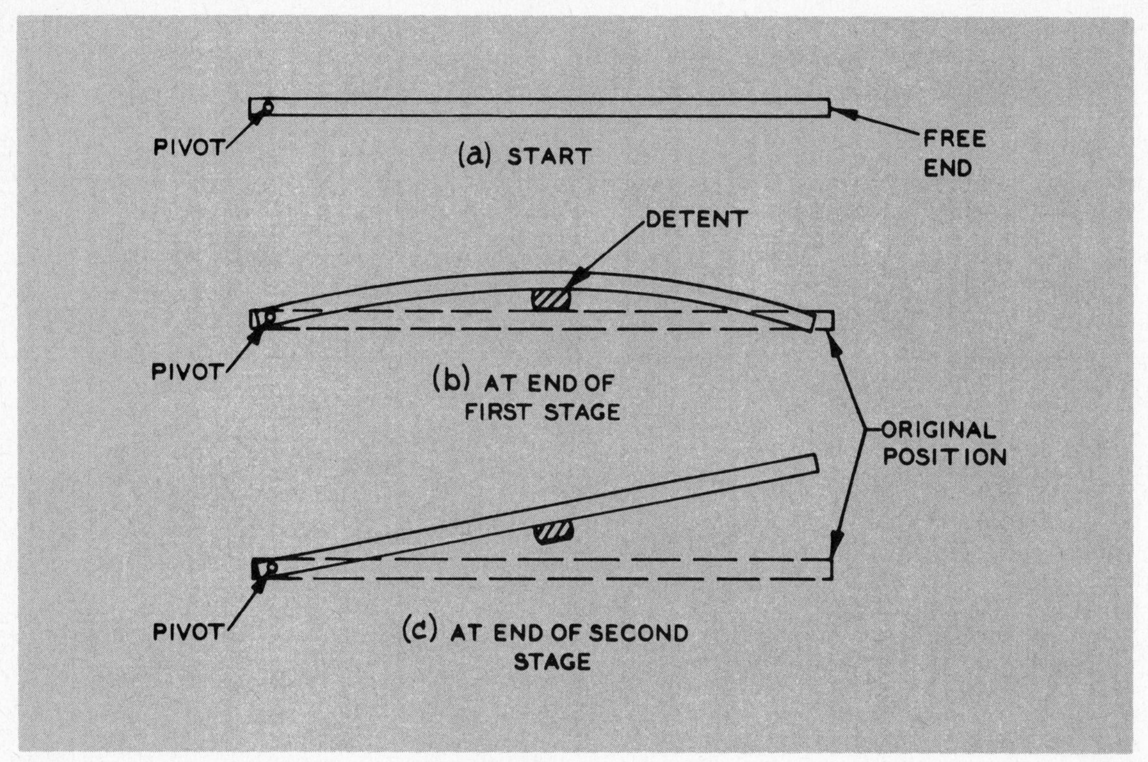

Fig. 5-14. Mechanical hookup of two-stage toaster.

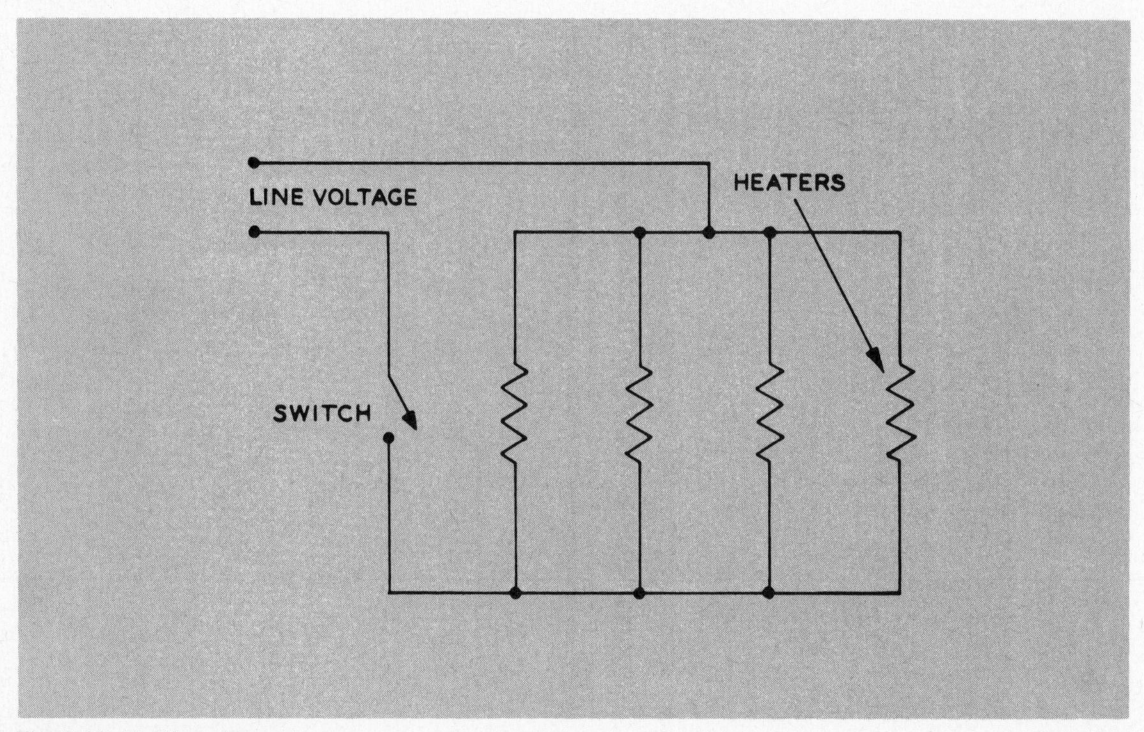

Fig. 5-15. Heating elements.

connected directly to the outlet. This could be the high side of the line and you are unlikely to know which side of the plug is connected to the hot side. Now if you touch a heating element and ground at the same time, you will be across the line voltage. When working on a toaster, always make sure it is disconnected except when you are making voltage tests.

In Figure 5-15, four heating elements are shown, which is usual for a two-slice toaster. These four elements are *not* identical, and thus if you have to replace one, make sure the new element has the same power or current rating as the old one. For example, in a 1200-watt toaster, the two outside elements may be rated at 315 watts each, and each of the inner ones at 285 watts. When slices of bread are in the toaster, the outer surfaces then are exposed to 315 watts. Each inner surface receives 285 watts from its adjacent element, but it also "sees" the other inner element and receives some heat from that. In practice, it receives about 10 per cent of the heat from the other element, so that in this example it would receive about 314 watts on each inner surface. Thus, all four surfaces receive about the same amount of heat. When ordering replacement elements, you can specify either current rating or power rating, but the best thing is to specify the model number of the toaster and whether it is an inside element.

If an automatic toaster doesn't work at all, fixing it is relatively simple. The electrical portion consists of a line cord, a switch, and the heating elements. Since it is unlikely that all four heating elements would burn out simultaneously, the trouble must be in the outlet, the line cord or the switch. The fault might also be due to an open circuit caused by a broken wire or a loose connection. First make sure that the outlet is live by testing for voltage. With the toaster disconnected from the line, remove the cover. The control knob usually slips right off since it is held on by friction. The handle which moves the lever is usually held on by a setscrew. The control knob and lever handle should be removed before removing the screws which hold the cover. It may also be necessary to remove some decorative trim.

With the cover removed, set the toaster right side up and plug it in. Check the toaster side of the line cord for voltage. If the line cord is bad, fix it or replace it, using the techniques discussed in Chapter 3.

If the line cord is all right, make a visual check of the toaster for loose connections – which are easily fixed – or broken wires, which can be fixed (at least temporarily) with a wire connector. Assuming everything appears normal, depress the carriage, and make a voltage check at both sides of the switch. You might also remove the plug, and with the carriage depressed, check the switch for continuity. A faulty switch should be replaced by an identical part.

If one of the heating elements fails to light, it may be open, but also it may only be disconnected. This is simple to check. A faulty heating element should be replaced with an identical part.

If the electrical portion of the toaster operates satisfactorily, but it won't pop up, or it pops up too soon, or otherwise acts improperly, the trouble is mechanical. Frequently, mechanical troubles can be cleared up simply by cleaning the crumbs out of the toaster. Many of the levers and linkages in the toaster depend on gravity for their action and require loose, clean joints to operate. Crumbs in a joint can interfere with proper operation. If necessary, put a drop of light machine oil on any sticky joint.

If the auxiliary coil on the bimetallic arm burns out, the carriage will not pop up when the toast is done. Check this heater for continuity.If it is open, replace the coil, if it can be removed, or the whole bimetallic assembly otherwise.

If toast pops out of the toaster, the shock absorber is probably defective. It should be replaced. A temporary repair can be made by stretching the spring that lifts the carriage so that it doesn't pull with as great a force. This is not recommended, however, since then the spring will not have enough force after the shock absorber is replaced.

To replace a heating element, it may be necessary to straighten the small crimped ears that hold it in place. This must be done carefully with long-nosed pliers; make sure that each ear is moved only enough to remove the old element. Fortunately this is something you will not have to do often, for too frequent bending would cause the ears to break off.

Before attempting extensive repair work, note carefully how the bread guide wires are mounted in the toaster. These thin wires keep the bread slices in the proper positions relative to the heating elements, but because they are easily bent, they can be damaged by careless handling. If you observe how they are held in the toaster, you can replace or straighten them correctly when they are damaged.

Ornamental trim has no effect on the operation of the toaster, but can be damaged if the toaster is dropped. (Internal parts also will be damaged.) The trim is easily replaced since it is usually held on by screws in an obvious manner.

When you have finished repairing a toaster, always make a ground check before returning it to service. Make sure no voltage exists between the case and ground regardless of the position of the plug in the outlet.

5-11. Rotisseries

A rotisserie or roaster is essentially a portable oven and in principle of operation is identical with the large wall ovens used in many households. It is basically an enclosed box with a heating element and a revolving spit, but more deluxe models come with many additional features. These include a separate baking element, a surface heater and a warming compartment. Some models with only one heating element may be turned upside down so that they can broil food when the heating element is at the top or bake food when it is at the bottom. Many models have an auxiliary timer which shuts off the rotisserie at a preset time. Some also include a timer which turns on the switch at a preset time as well as turning it off.

The broiling element in a rotisserie is usually a Calrod heater mounted in the top of the heating chamber, although coiled nichrome wire is also used. If a separate baking element is available, this is usually removable and may be plugged into a socket near the floor of the rotisserie. The baking element also may be Calrod or nichrome coils. The heating circuit usually includes, besides the heating element and line cord, a switch, a pilot light and some sort of heat control such as a thermostat.

The spit is rotated by a small electric motor. A separate switch is provided to run this motor, so that the spit may or may not be rotated when the broiler is used. The spit is also removable, and when it is out, as for baking or roasting, the motor can remain turned off. However, in some models, the motor runs constantly and, through a cam, controls the temperature both when it is driving the spit and when the spit is removed.

The main problem in servicing a rotisserie is in isolating the circuit you wish to test. Thus, you might make a continuity check across the prongs of the line cord, but you would be checking several circuits in parallel, and if any one of these is all right, you would have continuity. The motor circuit, the broiling element and the baking element must be checked separately.

A little thought will indicate what part of the circuit is at fault. If nothing works, the trouble is in the line cord, the main switch or the house circuits. Since a rotisserie draws more current than the average appliance, especially when both heating elements and the motor are turned on simultaneously, it may easily overload the house circuit, when other electrical appliances are also in operation. Hence, if a rotisserie doesn't work at all, check the outlet first. If the outlet is live, check the line cord, using the techniques described in Chapter 3, and if the line cord is at fault, fix it by the methods discussed in that chapter. Check the switch for continuity. Replace it if necessary with a similar one or one which has an equivalent current rating.

If everything works but the pilot light, you might ignore the trouble, and at your leisure replace the bulb at some later date. The bulb has no effect on the operation of the rotisserie.

If the motor turns, but either heating element fails to come on, the trouble is not in the line cord. It may be in the heating element itself or it may be due to a defective heat control unit. Check the heating element for continuity, making sure nothing is connected in parallel with it. If the heat control unit is a thermostat, make a continuity check across it, too. It may be stuck in the open position. Also check the connections to the heating elements. Sometimes, especially on plug-in elements, the contact pins oxidize, and this layer of metal oxide acts as an insulator. The contact pins should be cleaned with sandpaper. Never use emery paper to clean electric contacts, since the dust from this paper is conductive, and if it falls in the wrong place, it can cause a short circuit.

If the elements heat up, but the motor doesn't turn, check the motor switch. If the switch is all right and there is voltage across the motor terminals, the trouble is in the motor. The operation of electric motors and repairs you can make are discussed fully in the next chapter.

In one type of rotisserie, heat control is maintained by a cam on the motor. The cam opens and closes an auxiliary switch to the broiling element. When the control knob is set at *high*, this switch is out of reach of the cam, and the heating element is on all the time. At *medium*, the switch is pushed closer, and the cam opens it half the time. Thus, the heating element is drawing current only half the time and gives off half as much heat. The control knob at lower and intermediate settings moves the switch to an appropriate position so that the cam can open and close it to produce the desired heat. This type of heat control is used only on the broiler. An additional heat control in the form of a thermostat may be used for baking or roasting. If the cam slips on its axle, the switch may be stuck in the open position, or the cam may never open it, so that it remains closed and gives too much heat. A slipping cam is easily detected by observation and easily corrected by tightening the setscrew holding the cam to the axle.

After repairing a rotisserie, make a ground check before using it again. There must be no voltage on the case of the rotisserie with the line cord plugged into the outlet in any position.

5-12. Heaters

Electric heaters come in a wide assortment of shapes and sizes. Small, portable units are used for local heating or auxiliary heating. Larger units may be installed permanently, in baseboards for example, as prime heaters. Although circuits differ somewhat, the principles of operation of all are similar.

The heating circuit consists of one or more heating elements, a thermostat, selection switches to choose the desired amount of heat and a tip-over switch. In addition, there is a fan to blow the heated air into the room to be heated. The fan is operated by a small AC motor.

A simple circuit for a heater is shown in Figure 5-16. When the plug is inserted in a wall receptacle, the line voltage is applied across both the heating element and the motor through the tip-over switch and the thermostat. The temperature at which the thermostat closes is controlled by a knob on the front of the heater. In some models there is no on-off switch, and with the thermostat at its lowest setting the heater will be turned on by the thermostat whenever the ambient temperature (temperature in the vicinity) drops to about 55 degrees. To make sure the heater remains off when it is not in use, the wall plug should be disconnected.

A tip-over switch is an important part of every portable heater. This is a push button switch on the bottom of the heater which must be held in for the heater to operate. With the heater in its normal, upright position, the tip-over switch is pushed in because it is against the floor. If the heater is picked up or tipped over, a spring forces the push button out, disconnecting the electricity. The main purpose of this switch is to prevent fires.

The circuit of Figure 5-16 produces only one level of heat. When the room or space is sufficiently warmed, the thermostat shuts off the heater; and when the room cools down,

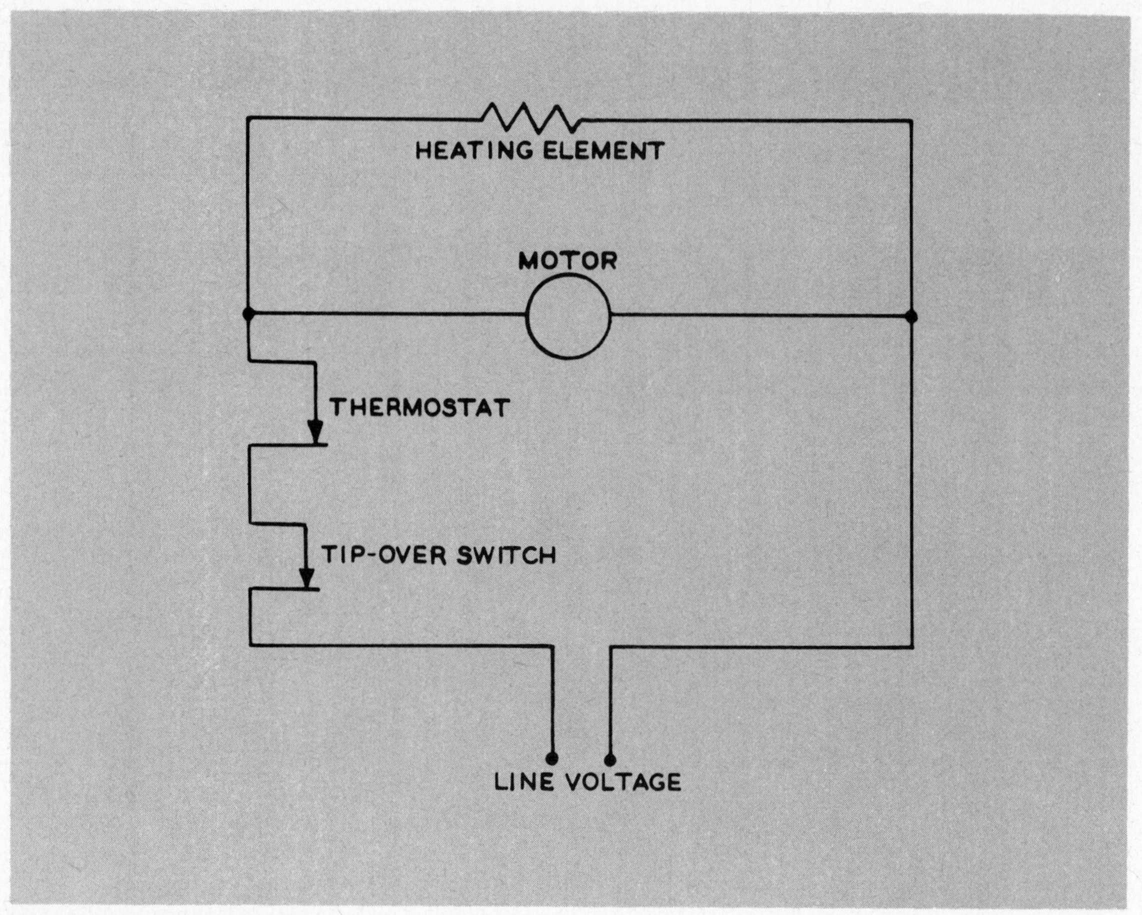

Fig. 5-16. Simple heater circuit.

the thermostat turns it on again. Other types of circuits are shown in Figures 5-17 and 5-18. In the heater shown in Figure 5-17, only one element is used, but the voltage across it is changed by a series of switches. Push button switches are shown in the figure, but any type could be used. When the high, medium or low button is pushed, the motor is also connected at the same time. The dotted line around the switch in the motor line indicates that it is not necessary for the user to close this switch since it is done automatically when any of the others are closed. When the *high* switch is closed, the full line voltage appears across the heating element, producing maximum heat. When the *low* switch is closed, two resistors are placed in series with the heating element. The voltage across the heating element is reduced, since there is a voltage drop across the resistors. This causes a lower current through the heating element and consequently lower power or less heat. At the *medium* setting, only one of the voltage dropping resistors is in the circuit, so that the heat produced is intermediate between the other two. The push buttons are arranged so that pushing down one releases all others. An additional button is used for off, which disconnects all the rest, including the motor switch.

The safety thermostat shown in Figures 5-17 and 5-18 is not used to control the operation of the heater as was the thermostat in Figure 5-16. Instead this is strictly a safety device which shuts off the heater completely whenever the temperature gets dangerously high.

The heater depicted in Figure 5-18 uses two heating elements. Typically, one may be

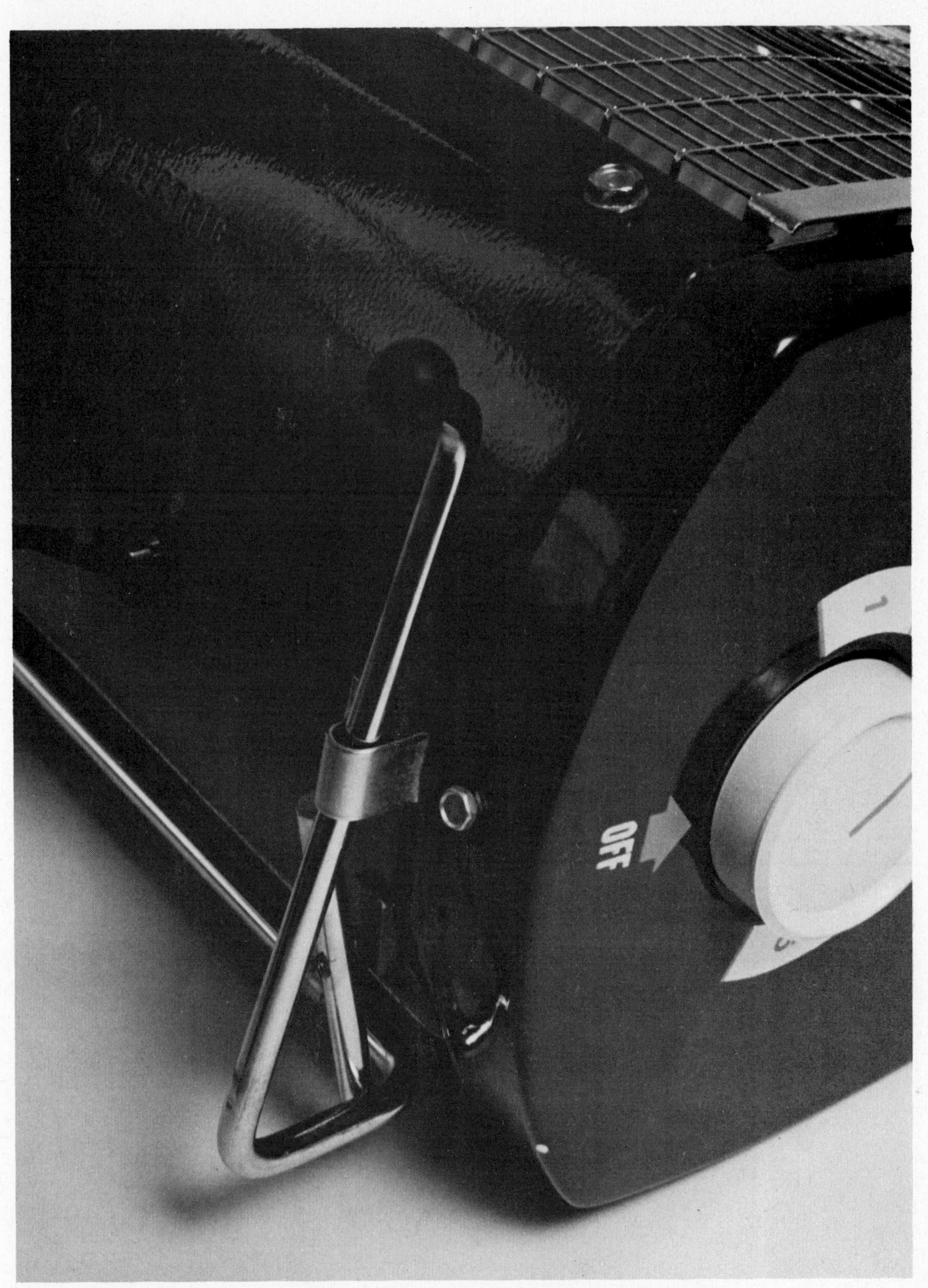

Electric heaters have a tip-over switch on the bottom that breaks the circuit if they accidentally fall over.

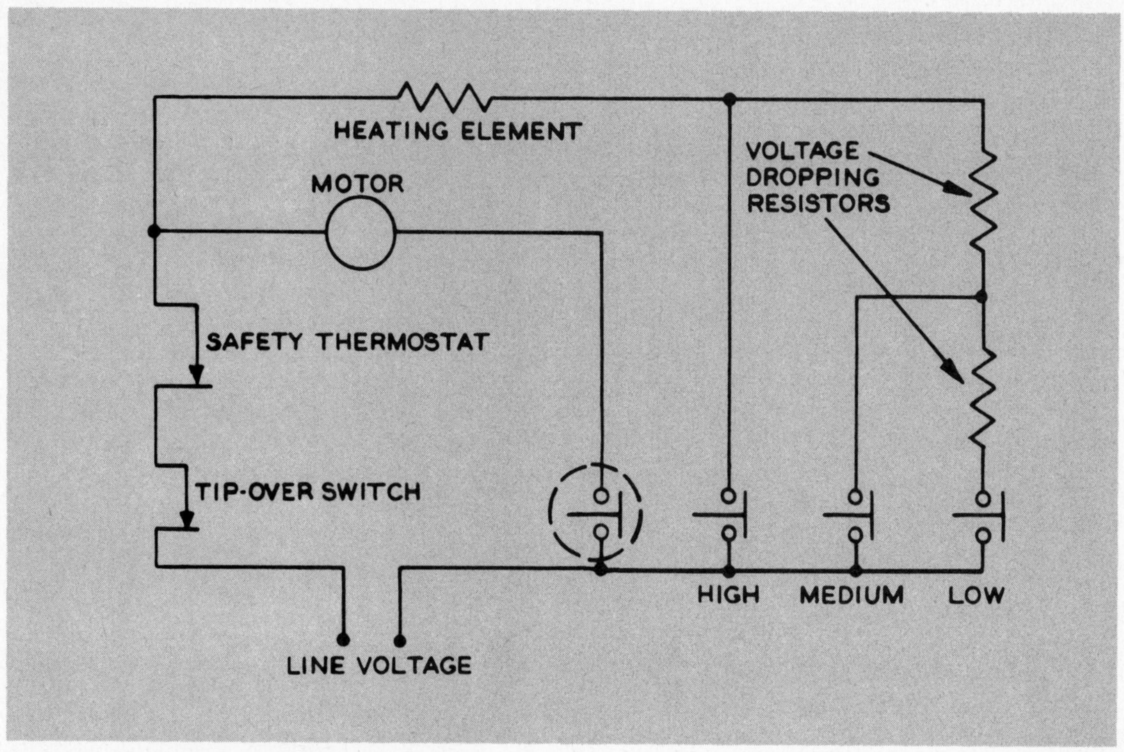

Fig. 5-17. Push button variable heat control.

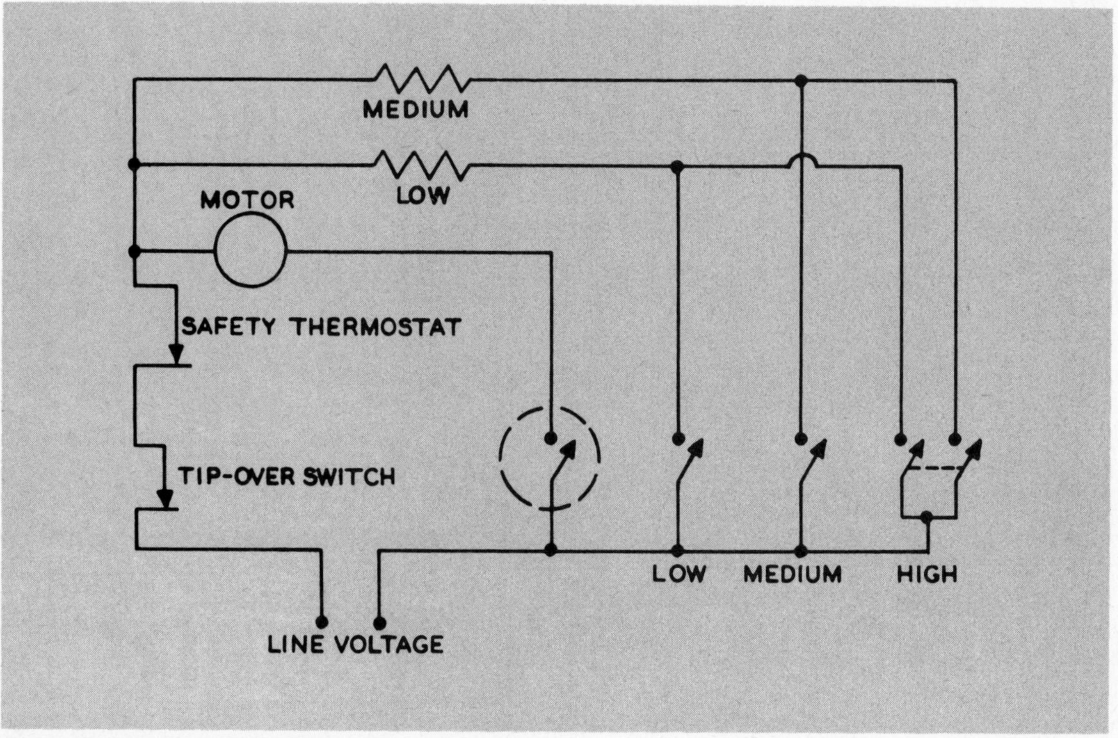

Fig. 5-18. Two-element heater.

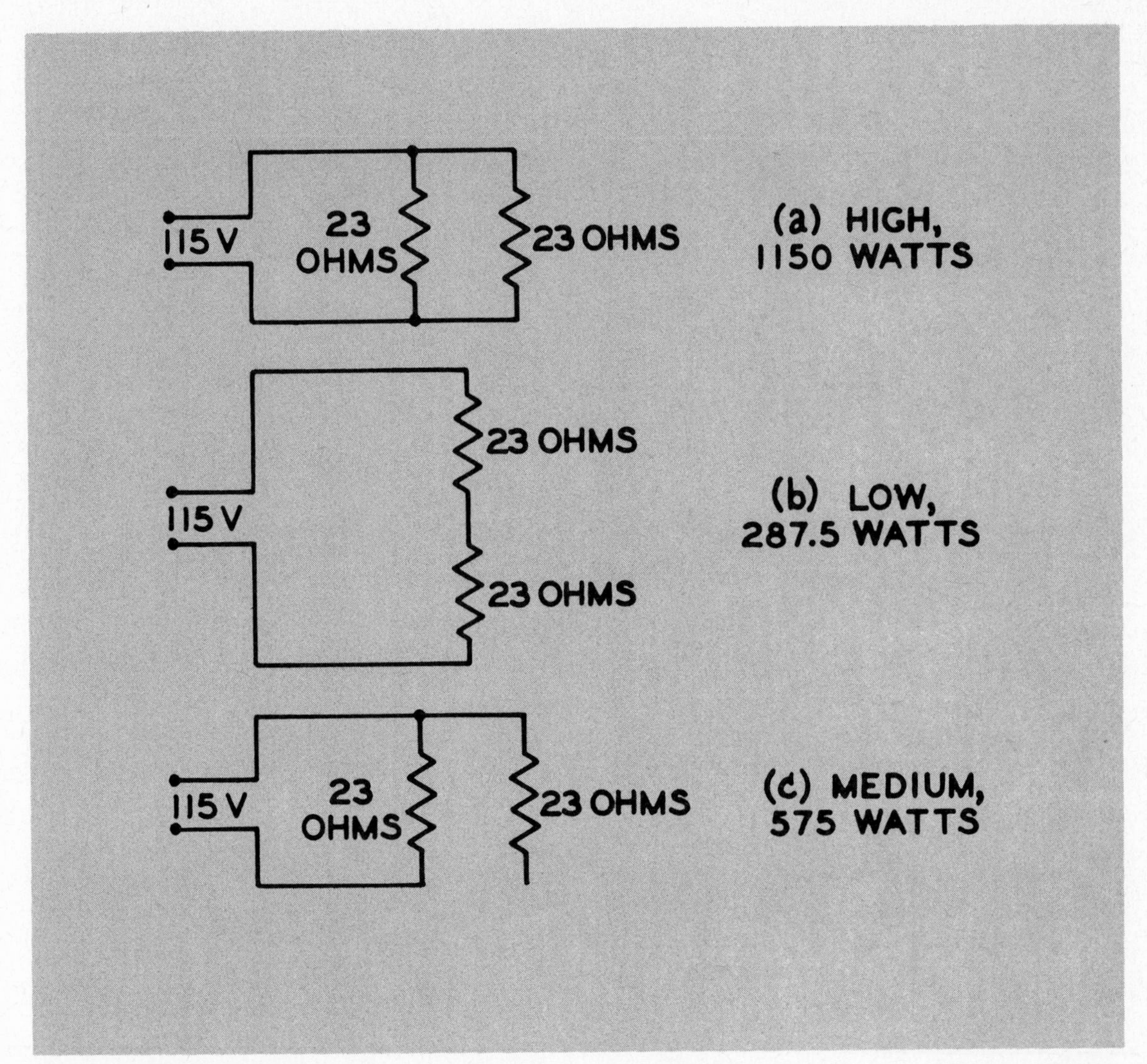

Fig. 5-19. Heating circuits.

rated 400 watts and the other 800 watts. Separate switches are shown, although one multi-purpose rotary switch may be used. When the switch is set for low, only the 400-watt unit is in the circuit; at *medium*, only the 800-watt unit is connected; and at *high*, both are connected, producing 1200 watts. Thus, three levels of heat are available.

It is also possible to produce three levels of heat using two identical elements. A typical circuit is shown in Figure 5-19. Two elements, each with a resistance of 23 ohms, are used. When the switch is set to *high,* both elements are connected in parallel, as in (a). Since there are 115 volts across each element, the current through each is five amperes. (Recall that voltage equals current times resistance.) Thus, the total current is 10 amperes, and the power is current times voltage, or 1150 watts. When the switch is set for *low,* the elements are connected in series, as in (b). Now there are 46 ohms across the 115-volt line. The current is 115/46 or 2.5 amperes. The power is, therefore, 115 times 2.5 or 287.5 watts. In the *medium* position only one element is across the line, as in (c). The other is disconnected. The current through this element is five amperes, and the power is thus 5

A simple way of improving the efficiency of a heater is to shine the reflector.

times 115 or 575 watts. This type of circuit is used in some rotisseries as well as in some electric heaters.

If neither the fan nor the heating element operates when the heater is plugged into a live outlet, the trouble may be in the line cord, the thermostat, one of the switches or a loose connection. Check the line cord first. Usually, the back of the heater can be removed by loosening four screws, and this permits access to the electric circuit. If there is no voltage at the heater end of the line cord, repair or replace the line cord as described in Chapter 3. If the line cord is in good shape, as indicated by voltage at the heater, make continuity checks on the switches and thermostats. When checking the tip-over switch, you must make sure the switch is depressed when you check for continuity. You can short out the switches and thermostat temporarily with clip-leads. Now the heater should operate if one of these is bad. Remove one clip-lead at a time, and when you have removed the lead across the faulty component, the heater will no longer work. Be sure you pull out the plug whenever you put on or remove a clip-lead.

If one or more heating elements fail to work, although the fan motor runs, the trouble may be in the element or the switch connected to it. In either case, a continuity check will identify the defective part. Defective switches and thermostats should be replaced with parts with similar ratings, if possible, although slight deviations will not impair the operation of the heater. In some heaters, individual heating elements may be removed and replaced with new ones. In others, the heating elements and reflector are not separable, but must be replaced as a unit. These parts are generally available at authorized service dealers, but if not, they may be obtained from the factory at the address on the nameplate.

If a heater seems to work, but doesn't produce enough heat, one of the heating elements is probably defective. The trouble may also be in the switch connecting this element to the line. Continuity checks will enable you to locate this trouble. Or it could be simply that the reflector is dirty. If so, the guard is easily removed, and metal polish applied to the reflector and thoroughly rubbed off.

If the heating elements work, but the blower motor is inoperative, the trouble may be a defective motor; but it may also be mechanical, such as a bent fan blade jammed against the frame of the heater. First make sure that the motor and fan are free to rotate without obstruction. Make continuity checks on the switches which operate the motor, and also check to see that all connections are in order. If you decide that the motor is at fault, it may be possible to repair it, or if not, replace it. Motors are discussed in detail in Chapter 1 of Volume 10.

Finally, after repairing the heater, make a ground check to be sure there is no voltage on the case for any position of the plug in the socket.

5-13. Vaporizers

An electric vaporizer is one of the simplest and most efficient of all electrical heating appliances. All the electrical energy is converted into heat, and none of this heat is wasted. It is all used to heat water. This is accomplished by making the water to be heated a part of the electric circuit, shown in Figure 5-20. The line cord is connected to two electrodes which are close together, but not touching, at the bottom of an insulated bowl. When this bowl is empty, no current flows. However, when water is placed in the bowl, the water completes the circuit, and current flows through the water. Since water has some resistance, the electric current produces heat, just as it does in any resistance. But this heat is entirely in the water, so none is wasted. Note that when the water boils, the circuit of Figure 5-20 will shut itself off as soon as all the water is evaporated.

In a practical vaporizer there must be some way of replenishing the water supply as the water boils away. One method of doing this is shown schematically in Figure 5-21. A large jar acts as a water reservoir. It is filled with

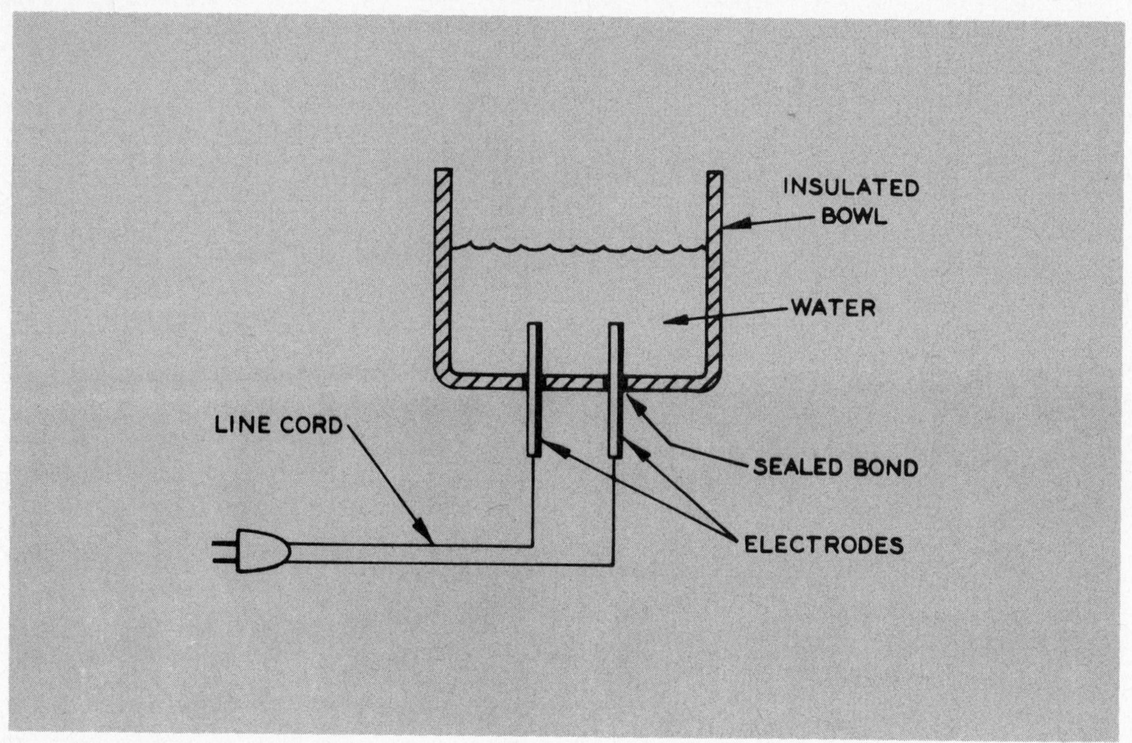

Fig. 5-20. Electric circuit of vaporizer.

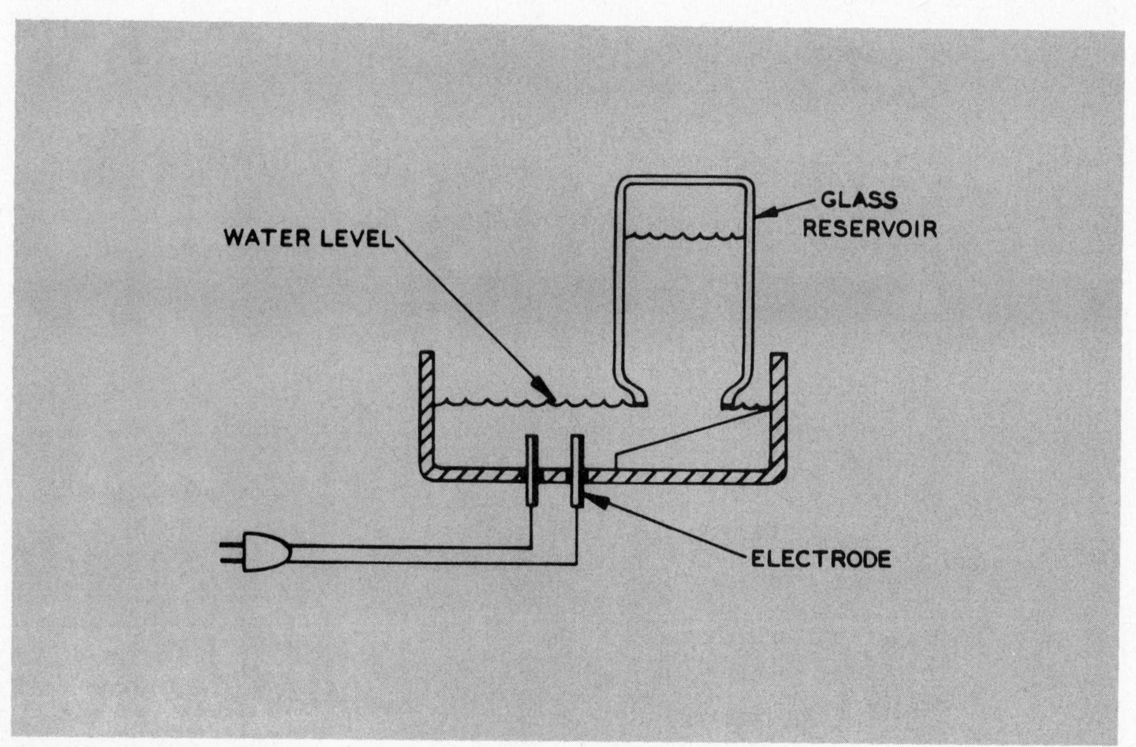

Fig. 5-21. Water reservoir.

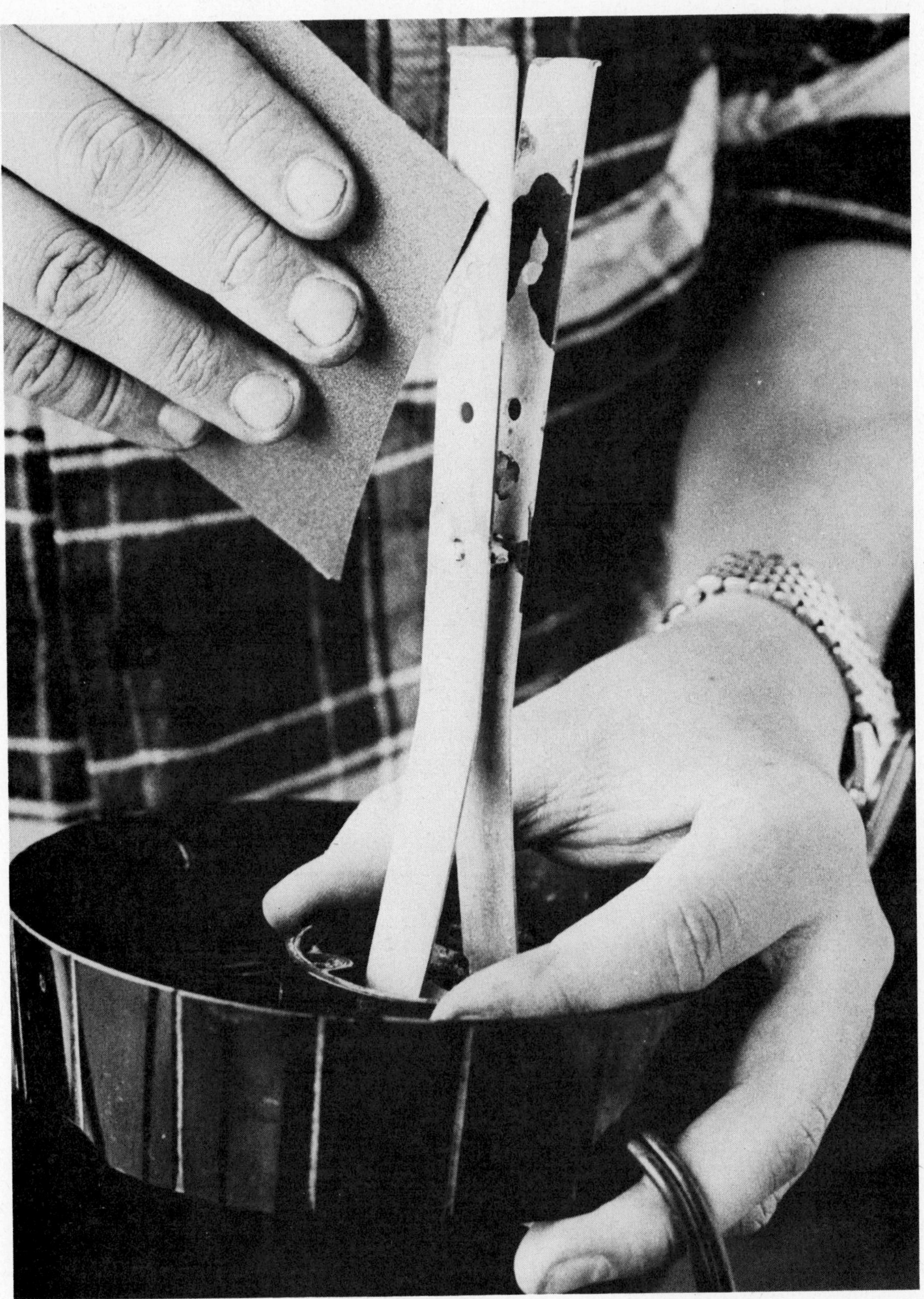

To avoid the short circuiting of a vaporizer due to the accumulation of mineral deposits, scrape the electrodes with sandpaper.

water and covered with a cap in which there is a small hole, about one quarter inch in diameter. This jar is inverted and supported in the insulated base so that the water will run out through the hole into the well containing the electrodes. When the water level in the bottom of the bowl reaches the cap of the jar, no more water will come out of the reservoir. This happens because the pressure of the atmosphere pushing on the water in the bowl is greater than the pressure of the column of water left in the jar. Now as the water is heated and evaporates, the level in the bowl goes down below the cap of the jar, and then more water comes out to keep the level constant. This continues until all the water is boiled away.

The electrical part of a vaporizer is amazingly simple, and almost nothing can go wrong with it. Check the line cord, and if that's all right, the circuit is all right. The problems that arise are not basically electrical in nature.

In some applications, medication is placed in the bottom of the bowl to be evaporated along with the water. If the medication is greasy, it can leave a coating on the electrodes, and when this coating is thick enough, it may act as an insulator to prevent the electrode from making electrical contact with the water. To prevent this or to correct it when it does happen, the vaporizer should be washed in soapy water after each use.

If the local water supply contains a high level of mineral salts, these may be deposited on the electrodes as water boils off. Usually, these minerals conduct electricity and do not interfere with the operation of the vaporizer. However, eventually they may build up to a point where they short circuit the electrodes, usually blowing a line fuse. To prevent this, you should scrape the mineral deposit off the electrodes from time to time.

A special application of this method of heating water is a warmer for baby's bottles. A measured amount of water is placed in the bowl containing the electrodes, and the bottle is also inserted above the water. The water is heated and boils, and the steam heats the bottle. Since the amount of water is controlled, the time for it to boil away is also controlled. This time is selected to bring the baby's bottle to just the right temperature.

Preventive Maintenance

The average home has about twenty to thirty electrical appliances. If each appliance had only one failure per year, you would expect then to have an average of a failure every other week. Clearly this is too many, but fortunately most appliances, even when abused, give good performance for much longer than a year. Nevertheless, when an appliance does fail, it is always inconvenient, and it is all the more painful because the failure probably could have been averted or at least delayed for many months by using proper operating procedures and suitable *preventive maintenance.*

If you have to call a repairman, you can expect to pay a bill which is large compared to the original cost of the appliance, even to replace a part costing only a few cents. In view of the high cost of repairs, it is truly amazing that more than one in ten service calls is to replace a burnt-out fuse, to tighten a loose fuse or to plug in an appliance because the user didn't realize the plug had been pulled out. If you are able to do your own appliance repair work, you will solve the fuse problem or plug in the appliance in situations like these, and you can then congratulate yourself on how much money you've saved. It takes little effort, however, to make sure that the fuses (or interlocks) are good in the first place, and that line cords of frequently used appliances remain plugged in. By the same token, the amount of extra effort needed to operate appliances correctly and to keep them in operating condition is trifling compared to the extended period of trouble-free operation which results.

One of the greatest causes of failure in an electric appliance is not a fault of the appliance at all, but a failure in the line cord or plug. After reading Chapter 3, Volume 4, you know what types of failure you might expect in line cords and plugs, and hopefully you now know how to fix them. However, almost all of these failures can be prevented. Here are some reasons why line cords fail:

1. The user yanks the plug out of the outlet by pulling on the cord.
2. The cord is bent sharply around an obstruction, or a piece of furniture is pushed against a plug, causing a sharp, right-angled bend in the cord.
3. The appliance is dangled by the cord.
4. The cord is run behind a radiator or along a window sill in bright sunlight.
5. The cord is run under a rug and is stepped on frequently.
6. Someone steps on a plug, cracking it.

This is by no means a complete list, but it

Electrical cords should never be sharply bent around an object of furniture.

Dangling an appliance by its cord will cause undue wear on the cord.

Extension cords should not be placed under a rug; the weight of people stepping on them will break or damage them.

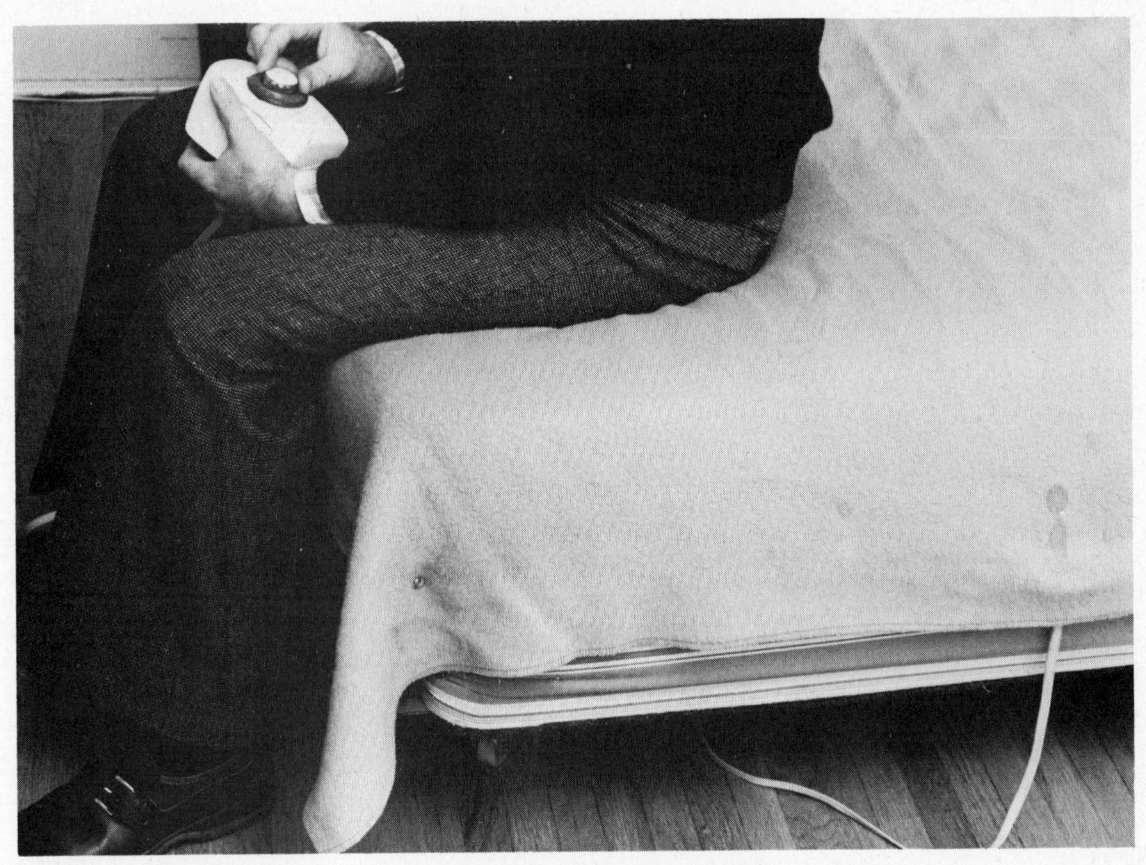

Sitting on an electric blanket may cause the heating elements to break.

should give you a good idea as to the causes of cord failure. Also, it is not difficult to figure out how most of these troubles can be prevented. The deceptive part of this problem is that even when the cord is abused frequently, the appliance continues to work well, so it seems hard to justify the extra effort needed, for example, to reach over and pull out the plug by grasping the plug itself instead of yanking the cord. Nevertheless, every time the cord is yanked it shortens the time to the next failure.

When you buy a new appliance, read the instructions before using it. As new improvements are designed into appliances, new methods of operation are needed. It is important for the user to know the limitations of an appliance as well as its advantages. Unfortunately, salesmen stress the advantages and gloss over the limitations. When you buy an appliance, it is best to follow the manufacturer's instructions to the letter. However, auxiliary supplies are also improved, and some changes may be in order. For example, if your dishwasher requires one teaspoon of detergent (according to the instructions supplied with the machine), it may be desirable to use only 1/2 teaspoon of a new, improved detergent.

An electric blanket should give many years of service when used according to the manufacturer's directions. The most common source of trouble in an electric blanket is overheating caused by leaving a weight on the blanket while it is on. The weight may be only a book which you intend to read in bed, but if it rests on the blanket too long, the heat under it cannot escape and may burn the insulation or the wire itself. Also, if an electric blanket is "tucked in", it is bunched or folded so that heat cannot escape. It should be obvious that you should never sit on an electric blanket, even when it is off. The heavy weight on the fragile wires may cause them to

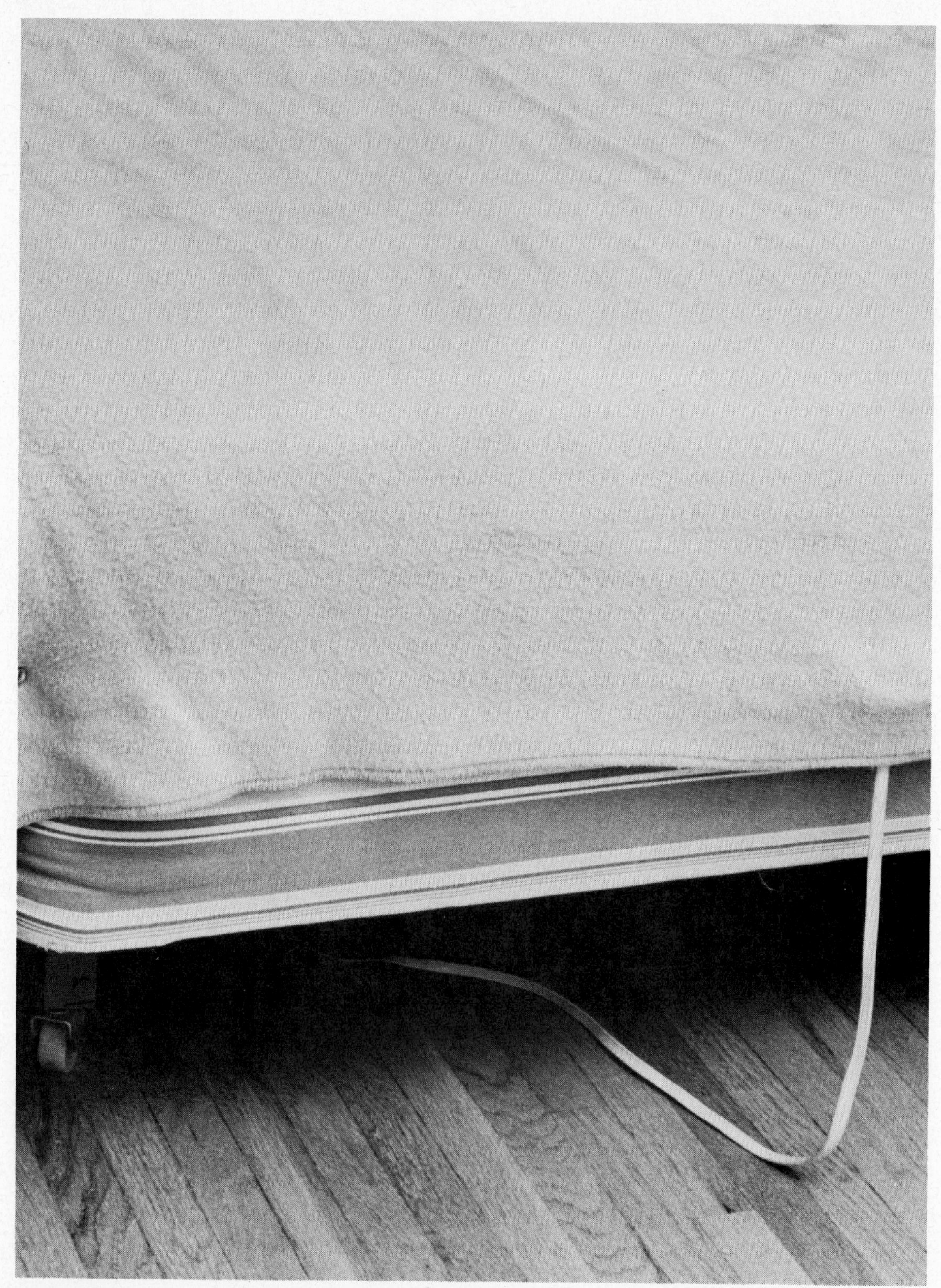

Let electric blankets hang freely over the edge of a bed rather than tuck them under the mattress; this avoids bending the heating elements.

break. In summary, electric blankets will give long life, if you remember these restrictions:

1. Don't sit on the blanket.

2. Don't put a weight of any kind on the blanket when it is on. This includes a bedspread over the blanket.

3. Don't tuck in the corners.

4. When the blanket is folded and put away, don't place any weight on top of it.

Where applicable, these same restrictions apply also to heating pads.

If a little is good, a lot is *not necessarily better. This is an important rule to keep in mind when using detergents in dishwashers and clothes washers. Too much detergent can harm an appliance, although the new low-suds detergents cause less trouble than the old sudsing varieties. If you have soft water in your pipes, you should probably use only half the amount of detergent that is recommended by the manufacturers. It is well to experiment when you first get your washer, and use as little detergent as possible to do a satisfactory job. Special rules for dishwashers are:*

1. Make sure the water is at or above the minimum temperature specified by the manufacturer.

2. Don't use the dishwasher as a garbage disposer. Dishes should be scraped first.

3. Learn how to clean the filter and keep it clean.

4. Make sure silverware cannot interfere with the operation of the dishwasher.

Garbage disposers are real work savers, but are abused frequently. They cannot dispose of everything. It goes without saying that spoons, forks, and other items of silverware should not be dropped into the disposer. Items which are not satisfactorily ground include fibrous materials such as corn husks, watermelon rinds and celery. Avoid raw bones. Although these would be ground up eventually, they put too heavy a load on the machine. If the disposer jams, don't panic. Use a long wooden stick to turn the turntable backwards and loosen whatever caused it to jam. You may have to press the restart button to start it again.

Toasters require some special attention. As bread is toasted, pieces frequently break off and fall inside the case. In most cases, small pieces and crumbs fall all the way through and may be removed through an access door. Larger pieces, however, can get caught in the body of the toaster and jam the mechanism or catch fire. It is important then to make sure the toaster is kept clean. The access door should be opened and all crumbs removed periodically. Once a month is not too often. Larger pieces should be removed whenever you know they are there. *Make sure the plug is disconnected whenever you clean out the toaster. Also, make sure you plug it in again when you finish cleaning it.*

A toaster probably takes more abuse than any other appliance. Never slam the handle down. This may bend vital parts, and although the toaster still works, its useful life is shortened.

Refrigerators and freezers should give many years of service. Some important steps to help achieve this are:

1. Make sure the condenser coils are kept clean by vacuuming them at least every six months.

2. Do not keep the refrigerator or freezer in a warm place, as it will have to work harder.

3. Don't pile things around or on top of the unit, as it works better when air circulates freely around it.

4. When the weather is warm or when an extra amount of food is in the cabinet, turn up the coldness control.

5. Check the door gaskets periodically and

When cleaning a toaster or doing any electrical work on an appliance, make sure the plug is in plain view so you do not mistakenly assume it is unplugged.

change them before they become brittle with age.

Motors in many appliances are sealed, self-lubricating units and need no maintenance. In others they should be oiled once or twice a year. As a rule of thumb, if there is an oil cup on a motor or blower or other moving part, you should oil it. But don't flood it with oil. Two or three drops each time are sufficient.

Dirt causes problems in all electric appliances. It was already mentioned that the condenser on a refrigerator should be vacuumed periodically. Similarly, clothes dryers, air conditioners and other large appliances should be checked for dirt and lint. It is a good idea to remove the sides of the cabinet and vacuum out any dirt you can see. Also, filters should be cleaned or changed according to the manufacturer's instructions. For air conditioners and gas furnaces, filters are cheap and should be changed about once a month.

Vacuum cleaners are used to suck up dirt and dust, and consequently, it is surprising that they, too, must be kept clean. Usually, some lint or dirt remains in the hose when the vacuum cleaner is put away. After a few months the lint in the hose limits the amount of air that can enter. One way to prevent lint from remaining in the hose is to run the vacuum cleaner, sucking in clean air, for a few minutes after finishing each day's cleaning. This will cause the lint in the hose to be sucked into the bag. In some cleaners, it is possible to attach the hose to the blower end of the machine, and actually blow out the lint.

With some attention to preventive maintenance, appliances will last for many years. In fact, you may reach a point where you hope an appliance fails so that you can buy a new, improved model. Don't despair because your appliances last too long. If the new model is that attractive, you can donate the old one to a PTA rummage sale, and feel virtuous when you buy the new one.

If absolutely necessary, a fork may be used to extract toast from a toaster, but first disconnect the plug and be careful not to damage the elements.

Some safety hints for the home

Electricity is our servant. It does our heating, lighting, cleaning and cooking. However, abuse and poor maintenance of electrical equipment contribute to accidents. Reduce the risk of receiving an electric shock:

- don't use portable electrical equipment in your bathroom except an approved electric shaver or electric toothbrush, which are of special construction
- avoid handling appliances which are not grounded, especially when they are near sinks, water pipes or radiators unless the cord is disconnected from the conventional outlet. Provision should be made in your wiring system for grounding of appliances. If this is not done and a grounding conductor is not installed in your wiring system, and you simultaneously touch a faulty appliance and a metallic object (water pipes, sinks, etc.) then you could receive an electric shock
- don't overload circuits by using multi-outlet devices
- don't use extension cords to supply permanent equipment...appliances should be either permanently connected to or plugged directly into convenience receptacles. Don't leave unused extension cords plugged in. A child may put the end in its mouth
- don't replace blown fuses with fuses of a higher rating, or resort to any other means to restore power other than a proper fuse. Should the fuse continue to blow, call a reliable electrician
- most appliances, especially automatic washers, dryers, stoves and air-conditioners should be grounded, besides any electrical equipment used outdoors or in your basement. Make sure you have three-wire receptacles installed in your wiring system...use three-wire attachment plug caps on portable equipment
- don't allow children to poke objects (pins, or scissors) into receptacles...plastic safety covers should be installed if the children persist
- portable type heaters should be kept away from combustible material, and in a place where they cannot be knocked over...some models automatically shut-off if they are accidentally moved
- read and follow instructions on heating blankets and pads...avoid the use of pins and do not abuse by excessively stretching the blanket or pad
- make sure the insulating link is in its proper place on all pull-chain lighting fixtures
- keep all electrical equipment and appliances in a state of good repair
- disconnect all appliances such as irons, heaters and toasters when not in use

Make sure all major appliances and electrical equipment are grounded with a three-prong receptacle, unless approved with a two-wire cord.

Fuses and circuit breakers should not exceed 15 amperes except for heavy-duty appliances.

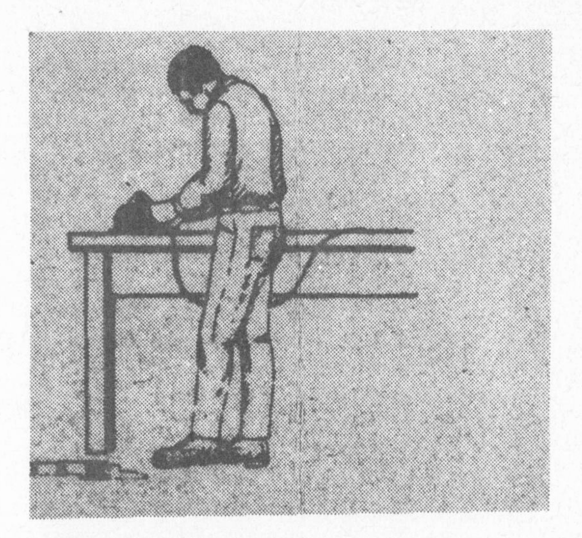

Avoid standing on a wet floor when using electrical appliances.

Pull main power switch before working on electrical circuits or changing a fuse.

Frayed cords or damaged appliances should be replaced or repaired.

For instruction on how to repair major electrical appliances, see Volume 10 of this series.

see Volume 10 of this series.

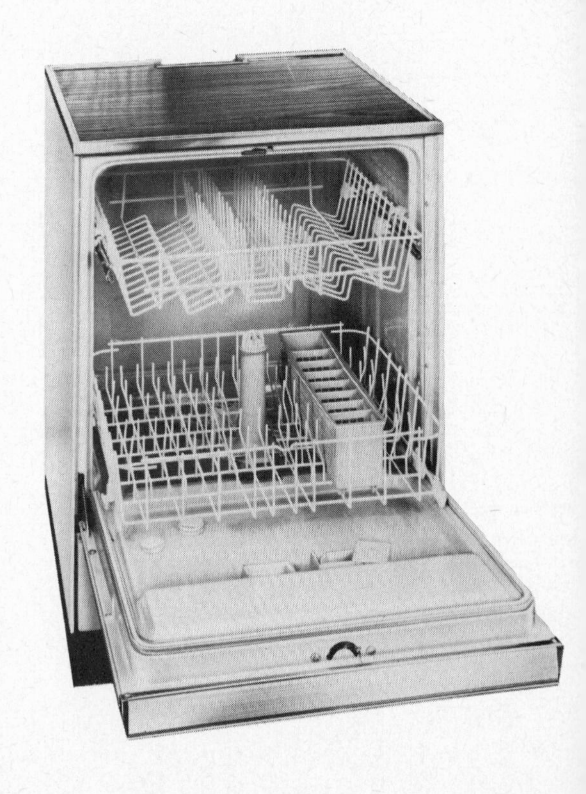